Female Tech Managers' Diversity
Perceptions

Larry J. Chamber

Abstract

This study explored female managers' perceptions of diversity hiring. Most diversity studies have used quantitative methodologies and have not explored the qualitative perceptions of diversity hiring. The study's three research questions ask which personal attributes of employees should be diverse within an organization, for which jobs should diversity be a criterion in the employee selection process, and in which organizational situations should diversity be a criterion in the employee selection process. The study follows the qualitative methodology of generic qualitative inquiry and included in-depth interviews of twelve participants. The population for the study is female managers who work in technology companies, within the United States. Analysis of the data collected from interviews, observations, and artifact review occurred through open, axial, and selective coding. The study finds that these participants value diversity among a wide variety of employee personal attributes, jobs, and organizational situations. The implications for practice are that hiring practices can promote diversity when they reduce bias, make diversity a priority throughout the recruiting process, and use targeted sources to reach a wider spectrum of candidates.

This book is dedicated to all of my family members for their support of my academic journey. Specifically, I thank Katya for her immense patience and flexibility during my five years of rigorous study. She even let me sell the house and insurance agency so we could afford to live in Europe while I studied full time. I am grateful to Yana and Akim for listening to my inelegant explanations of philosophical concepts. Hopefully you don't think Papa has gone completely crazy. My parents are appreciated for fostering an interest in higher education that dates back to my teenage years. There was never any question of whether I would go to college, just when and where.

Acknowledgments

I am grateful for the support of several key academicians who have provided insight and encouragement during my doctoral studies. These include my immensely insightful mentor and committee chair, Dr. Michael Williams; my book committee members, Dr. Elizabeth Rescigno and Dr. Angela Jiles-Charles; and inspirational faculty members Dr. Jane Gibson, Dr. Kathryn Kleypas, Dr. Pamelyn Witteman, and Dr. Bruce Chapman. Special thanks are due to fellow recent graduates, Dr. Sterlena Taylor, Dr. Jessica Kensington, and Dr. Thomas Marben, for guiding the path through the book forest. In addition, I thank Capella University and Hartnell College for their financial scholarships and grants.

Special appreciation and recognition are due for Arika D. Verma. Arika, your support and encouragement enabled me to overcome barriers that temporarily blocked the progress of this study. You are a shining light and a dear friend.

Table of Contents

List of Tables

List of Figures

CHAPTER 1. INTRODUCTION

Decades of civil rights reforms in the United States of America have resulted in increased opportunities for women and minority populations in the workplace (Gardner & Ryan, 2020). Although organizational diversity has increased during this time, it has not always led to improved organizational performance (Thomas & Ely, 1996). What is more, female and minority employees still report harassment and discrimination, as evidenced by the recent *Black Lives Matter*, *Times Up*, and *Me Too* movements.

While the U.S. population becomes more diverse and equality is increasingly sought by women and minority groups, the debates surrounding both the benefits and challenges of diversity continue. Questions arise such as whether diversity is correlated with organizational performance. If so, what are the mediating factors? How should organizational diversity be achieved? What is the role of inclusivity in creating a diverse and successful company? As organizational leaders look to use the power of diversity in organizational development, what are the roles of leadership and diversity hiring practices?

These questions surrounding diversity are relevant for all industries. They are especially relevant for technology companies. Technology companies have extensive economic and social influences on the world today. However, even tech giants like Google have been accused of discriminatory behavior, evidenced by their alleged firing of prominent AI researcher Timnit Gebru (Allyn, 2020) and the company's 2022 gender discrimination settlement (Grant, 2022). What is possibly more striking, Silicon Valley reportedly birthed the dark enlightenment or neo-reactionary movement, the precursor of the alt-right movement (Aikin, 2019).

Despite tech's white-male-dominated environment (Wynn, 2020), prominent female leaders, such as Sheryl Sandberg and Marissa Mayer, have ascended to top positions in technology companies. How have transformational female technology managers experienced diversity hiring, from their unique perspectives? What can we learn from understanding their vantage points? Have they paved the way for other women to follow in their footsteps? How is diversity influencing the tech industry as a whole?

These questions demonstrate how little we know about diversity in the context of today's mighty tech companies. It is an environment where the talents of women and minorities have not been fully utilized (DuBow & Kaminsky, 2019). There is a disproportionately low number of female executives in technology firms, so women's valuable perspectives are underrepresented (McGee, 2018). Their insights might help these companies progress and benefit more fully from the power of diversity.

Addressing the lack of diversity within technology companies involves understanding how diversity is experienced within this environment. The unique perspectives of those women who occupy leadership roles in these firms could provide valuable insights. Their struggles to transform organizational culture and hiring practices are of special interest. This study therefore investigates female managers' perceptions of diversity hiring in technology companies, including the information technology and bio-technology industries. Understanding these perceptions could help advance the body of knowledge related to transformational leadership and diversity hiring practices in technology companies.

Thomas and Ely (1996) posited that organizational diversity can increase business performance, but it requires that the leaders of the organization are committed to diversity. Since diversity is an essential component of transformational leadership (Bass & Steidlmeier, 1999),

this leadership style could be useful in fostering an organizational culture where leaders are

committed to diversity. Achieving this culture also requires hiring practices that promote

organizational diversity (Montei et al., 1996). This study's findings might help organizations

refine their diversity hiring practices. The hope is that improved diversity hiring could result in

increased organizational success (Powell, 1998), possibly contributing to increased profitability

and employee well-being.

Background of the Study

This qualitative study explores the intersection of transformational leadership style and

diversity hiring practices, as experienced by female managers in technology companies. General

system theory provides the conceptual framework for analyzing the interplay of these constructs

and relevant theories. Such related theories include transformational leadership theory (von

Bertalanffy, 1957), stakeholder theory (Freeman, 1984), and person-organization fit theory

(Chatman, 1989). Research regarding transformational leadership and diversity hiring has been

extensive. However, these two constructs have been studied less frequently in combination. A

brief overview of the literature related to each construct follows.

Transformational Leadership

Transformational leadership style was proposed by Bass and Stogdill (1981),

incorporated into the full range leadership model in 1991 (Avolio & Bass, 2001), and endorsed,

as well as critiqued, by Yukl (1999). This approach to leadership aims to motivate and enable

workers to transcend personal interests in pursuit of organizational objectives (Eberly et al.,

2017; Chai et al., 2017; Jiang et al., 2017). The transformational leadership style has been found

3

to promote ethical leadership (Bedi et al., 2016) and contribute to organizational performance (Borgmann et al., 2016), including team and business unit performance (Lord et al., 2017).

The role of gender in relation to transformational leadership style is not fully understood (Dappa et al., 2019). However, the proportion of women in positions of power within U.S. organizations has increased significantly in recent decades (Hekman et al., 2017). Since leaders influence organizational culture and practices (Gleibs & Haslam, 2016) and there are now more women in positions of power, it would be useful to know female managers' perceptions of diversity hiring. London et al. (2019) found that leaders who are oriented toward collective goals promote equity and responsibility. It would follow logically that transformational leaders might positively influence equity and responsibility within their organizations, evidenced by workforce diversity and fair hiring practices. There appears to be a unique opportunity for female leaders to promote diversity within organizations; for example, correlations have been found between corporate board gender diversity and organizational performance (Conyon & He, 2017; Moreno-Gómez et al., 2018). In addition, research into gender diversity's influence on group decision making within organizations has found that increasing female representation can improve decision quality (Lamiraud & Vranceanu, 2018), limit risky decisions (Bogan et al., 2013), and encourage compromise between opposing parties (Nikolova & Lamberton, 2016).

Diversity Hiring

There is heightened interest in gender diversity research. This is not surprising, given the increasing proportion of women in positions of power, exemplified by the election of Kamala Harris as the first female vice president of the United States. The recent *Me Too* and *Time's Up* movements against sexual harassment are additional reminders of the growing prominence of women in society and the workplace, underlying the importance of the perspectives of female leaders. Despite the interest in gender diversity, there is minimal qualitative literature related to the perceptions of female managers regarding diversity hiring practices or impacts.

Authors writing more generally about diversity hiring's impact on organizations have found that diversity contributes to organizational effectiveness because of the complementary perspectives and approaches that it adds within organizations (Mazibuko & Govender, 2017). This includes the knowledge, abilities, and perspectives that previously underrepresented populations can contribute within their workplaces (Meric et al., 2015). As a result, diversity management has been found to enhance innovation, market access, employee retention, and organizational image (Wondrak & Segert, 2015). Diversity management includes the establishment and execution of hiring practices that promote diversity (Ohunakin et al., 2019).

The research literature on diversity hiring practices indicates that a favorable perception of diversity leads to a positive view of incorporating underrepresented groups (Nakui et al., 2011). Conversely, an unfavorable perception of diverse populations results in mistrust and avoidance of individuals from such groups (Quadlin, 2018; Houkamau & Boxall, 2015). As a result, leaders can influence workforce diversity by modeling appreciation for its benefits and fostering an inclusive culture (Afful, 2018). Another way that leaders can influence workforce diversity is to recruit a sufficient pool of diverse job candidates, so that enough employees from underrepresented groups can be identified to meet stated diversity objectives (Yu, 2018).

Hiring a diverse workforce does not automatically result in increased organizational performance or improved employee satisfaction, however. One contributing factor might be that employee perceptions of organizational diversity practices differ based on their backgrounds

(Kundu & Mor, 2017). For example, women who presume that their organizations are not discriminatory sometimes do not recognize existing gender discrimination (Brady et al., 2015). Addressing such assumptions might allow organizations to further benefit from the synergy of a diverse and harmonious workforce. Exploring the perceptions of the female managers who are involved in diversity hiring will hopefully provide insights that managers and recruiters can implement, to enhance diversity hiring.

Rationale

The relevant literature has not thoroughly explained how leaders can increase diversity (Thomas & Ely, 1996) through hiring practices. Also, advocates of diversity hiring have not agreed on what priority diversity should receive over other hiring criteria, such as organizational fit (Powell, 1998). Therefore, the study seeks to explore female managers' perceptions of diversity hiring. Limiting the sample to female leaders allows insights that might differ from those of male leaders, who comprise the majority within the leadership of technology companies.

Purpose of the Study

There is minimal qualitative research in this area, as most diversity studies have used quantitative methodologies and have not explored the perceptions of diversity hiring. This research aims to address that gap. The knowledge gained might help organizations and their members understand how diversity hiring practices can benefit organizational stakeholders, which is not thoroughly understood (Thomas & Ely, 1996). Since qualitative studies use data that cannot be measured and yield results that cannot necessarily be generalized to a broader population (Latham, 2016), this study was designed to let the data tell the story. The results are available for scholars and practitioners to apply as they see fit. The application of the study's results will ideally lead to new insights in both future research and practice.

Statement of the Problem

The purpose of the study is to explore female managers' perceptions of diversity hiring. This understanding will hopefully help scholars and practitioners develop and implement strategies to enhance these leaders' appreciation of the benefits of workforce diversity (Kundu & Mor, 2017; Brady et al., 2015). An increased appreciation of these benefits could potentially lead to increased workforce diversity, which can result in a greater variety of perspectives and talents within their organizations (Chang et al., 2019). The increased availability of these perspectives and talents might improve organizational performance (Ohunakin et al., 2019).

Selecting and developing leaders with diversity-valuing traits can result in a more diverse and equitable organization. This is because leaders and their organizations benefit from minimizing the traits that act as barriers to equity (Frost, 2018; Ohunakin et al., 2019). Equity promotes diversity and inclusion, which can positively influence organizational performance. However, there are many factors that influence perceptions of diversity and thereby influence outcomes (Daniels et al., 2017; Kulkarni et al., 2018). Understanding how diversity is perceived among leaders is essential in crafting the hiring practices that are necessary to promote a culture of diversity (Quadlin, 2018; Afful, 2018; Yu, 2018; Menezes & Prikladnicki, 2018). It is on this foundation of prior research that an exploration of diversity perceptions and their influence on hiring practices can be achieved.

Significance of the Study

This study involves female managers who are transformational leaders. Through interviews, their perceptions of diversity hiring have been explored. Transformational leadership theory is a major theory of leadership that has evolved to explain how a leader can motivate a worker to transcend their personal interests (Banks et al., 2016). There is an established link between transformational leadership and organizational performance (Arif & Akram, 2018); exploring transformational leadership style's relationship with the perceptions of female technology managers might generate insights into its role in enhancing diversity hiring.

The study's outcome could potentially relate to other theories, lead to the development of a new theory, or influence the study of culture, organization, or human interaction. It is also possible that additional practices, designs, or methods will arise from the study's findings. For female technology managers, the successful application of insights that arise from the study will hopefully result in heightened personal and organizational achievement.

Research Questions

The research questions for this study are

- RQ1: Which personal attributes of employees should be diverse within an organization?

- RQ2: For which jobs should diversity be a criterion in employee selection?

- RQ3: In which organizational situations should diversity be a criterion in employee selection?

Answering these research questions could help advance the study and practice of business management because increasing workforce diversity has been shown to enhance organizational performance (Thomas & Ely, 1996).

8

Definition of Terms

Definitions of the research questions' terms and constructs are provided here in alphabetical order, so that the meanings will be clearly understood. The definitions have been paraphrased for clarity and are based on the relevant academic literature.

Diversity is defined as variation in gender, ethnicity, generation, relation to organization, seniority, power level, and work expertise (Harjoto et al., 2015), as well as the multiplicity of views, knowledge, and approaches to work offered by members of varied identity groups (Thomas & Ely, 1996).

Diversity hiring is defined as the consideration of diversity as a selection factor when making hiring decisions (Montei et al., 1996).

Transformational leadership is defined as the act of motivating an employee to accomplish more than they had previously planned to accomplish (Bass, 1985a).

Data regarding transformational leadership and diversity hiring were gathered through interviews, observations, and artifacts collected during the study.

Research Design

The qualitative research methodology is appropriate for this study because it deals with human perceptions (Latham, 2016). Generic qualitative inquiry is known by various names, such as basic qualitative research; it is a general method of qualitative research that is used when other approaches are not appropriate for the study (Merriam & Tisdell, 2016). Ethnography, the Delphi method, case studies, and other qualitative research methods would not be relevant, so this study applies generic qualitative inquiry to explore diversity hiring from the point of view of the study's participants (Percy at al., 2015). In a qualitative study, the researcher serves as the

9

instrument, so their biases have been identified and distinguished from the input of the participants (Merriam & Tisdell, 2016). The study's goal is not to judge the participants' perspectives, but to describe and possibly understand their perceptions (Erlingsson & Brysiewicz, 2013).

The objective of this study is to comprehend the essence (Dowling, 2007) of the perceptions of the managers participating in the research. Multiple methodologies were available to accomplish this, including descriptive, hermeneutic, lifeworld, interpretative, and dialogal approaches (Finlay, 2012). However, this study takes a generic qualitative approach. It applies Powell's (1998) model of extended fit to analyze participant hiring practices and perspectives, in relation to organizational fit and diversity.

A variety of techniques have been used to obtain the study's data. These included semi-structured interviews, memoing of observations, constant comparison, and member checks. Memoing is a method of documentation that is very flexible, allowing the researcher to use the technique in accordance with the goals and nature of the study (Birks et al., 2008). These techniques provide the rigor necessary for a qualitative methodology that enables valuable data to be gathered and properly analyzed.

Assumptions and Limitations

The assumptions identified in the relevant literature were not consistent. For this study, some general methodological assumptions can be identified. The idealist ontological perspective is that minds construct perceptions of reality (Kluge, 2006), the epistemological view is that the researcher and study participant influence one another, and the axiological belief is that ethics should precede profits. The study's theoretical assumption is that organizations act as open systems (von Bertalanffy, 1957). It is this view of organizations as open systems that provides the conceptual framework through which this study's constructs and supporting literature are viewed.

Topic-specific assumptions have also been identified for this study. It was assumed that the study's participants would provide honest and thorough answers, both in response to interview questions and regarding whether they self-identified as transformational leaders. Another assumption was that participants would agree to be met over video-conferencing software, allowing observation and documentation of body language, gestures, and facial expressions, in addition to each participant's tone of voice. The interview recordings contain only audio data, however. A final assumption was that any researcher biases could be adequately identified and minimized, not only before and during, but also after the interviews.

A methodological limitation of the study is that it did not produce generalizable results, due to the subjectivist nature of qualitative research (Latham, 2016). There is similarly no identification of statistical correlation, since that can only be accomplished through quantitative studies. The study's results are limited to the insights that were generated from the input of the sample of twelve participants, all of whom are female managers in technology companies.

Organization of the Remainder of the Study

This chapter provided the study's background, rationale, and design. The remainder of the study consists of four chapters and two appendices, following the standard model for philosophy doctorate dissertations at Capella University. Chapter 2 reviews the relevant literature and the study's theoretical orientation. Chapter 3 presents the study's qualitative and exploratory methodology. This includes a discussion of the study's validity and reliability, data collection and data analysis methods, and ethical protocols. The data are analyzed in chapter 4,

with discussion and recommendations presented in chapter 5.

CHAPTER 2. LITERATURE REVIEW

This literature review summarizes the body of knowledge related to the topic of this book: female technology managers' perceptions of diversity hiring. The aim of the literature review is to provide context for this study and its findings, since even qualitative studies require an understanding of the relevant empirical research (Rocco & Plakhotnik, 2009). The review presents a description of the relevant theories and constructs, an overview of the literature, a synthesis of research findings, critiques of previous research methods, and the theoretical orientation for the study.

Methods of Searching

The sources used in this literature review were accessed online through the Capella University library. Databases and journals provided the bulk of the articles and books included in the review. Materials not available in this fashion were obtained through Google Scholar or interlibrary loans. ProQuest's ABI/Inform Collection and EBSCO Host's Business Source Complete are the databases that were used most frequently. The literature researched for the study included material related to organizational management, decision making, human resources, and ethics. Key terms included, but were not limited to *gender, women, diversity, transformational leadership, and hiring practices.*

Conceptual Framework for the Study

The literature regarding transformational leadership and diversity hiring practices is not dominated by any single theory or model. A plethora of theoretical frames and associated constructs were present in the reviewed literature. The body of knowledge that has been reviewed for this study draws on a wide variety of related theories, including attribution theory, decision making theories, leadership theories, pyschometric theories, social identity theory, sociotechnical theory, and stakeholder theory. Since the literature on leadership and diversity hiring includes various theoretical perspectives, the review should likewise include multiple theories (Palmatier et al., 2018) and offer a balanced perspective, explaining any contradictions in the literature as well as methodological and theoretical gaps (Galvan & Galvan, 2017) .

Due to the broad scope of the literature's theories, models, and associated constructs, a conceptual framework is used as the theoretical foundation for this book. A conceptual framework can synthesize multiple theoretical and conceptual perspectives (Imenda, 2014). This is common when a study uses a qualitative and inductive approach (Imenda, 2014), as is the case for this study.

General system theory (commonly referred to as systems theory) provides a unifying theoretical foundation that serves effectively as the conceptual framework for this study. The reason is that systems theory explains the functioning of organizations at their most basic level (von Bertalanffy, 1957) . It can thereby be viewed as a foundation for newer organizational theories related to transformational leadership and diversity hiring practices, depicted below.

Figure 1
Study's Constructs Unified by Systems Theory

Systems theory (von Bertalanffy, 1957) connects the constructs and conceptual framework through its model of the organization as a system comprised of subsystems (groups and individuals) and existing within a larger ecosystem (the external environment). Systems theory (von Bertalanffy, 1957) was adapted from the natural sciences to other fields and expanded by various authors. According to Forrester (2016), the basic structures of system dynamics are the same for both natural and social sciences. In the business field, organizations can be viewed as systems; employees and leaders are internal components, while customers and suppliers are part of the ecosystem within which a company operates (Senge, 2006). Studying the dynamics that operate within and between systems allows organizational theorists to understand more thoroughly the patterns that arise (Senge et al., 1992). Analyzing organizations as systems can lead to insights regarding the highly complex nature of social interactions in a business environment, which arise from diverse values, interests, and perspectives (Scharmer & Kaeufer, 2010).

Organizational systems can evolve through continuous and collective learning (Schein, 2002). According to Senge (2006), learning more quickly than one's competitors is the only way to maintain an advantage over the competition. Systems thinking is necessary to accomplish this, as it integrates all of the disciplines of a learning organization (Senge, 2006). Forrester (1994) explained that interest in systems thinking is not sufficient, however. He cautioned against management fads and recommended a thorough understanding of system dynamics to apply systems principles properly. Forrester (2016) further explained that an understanding of systems allows one to recognize the underlying dynamics, regardless of other factors that might influence a system as it learns.

Systems theory relates to each of the theories that emerged in the literature review, viewed through the study's conceptual framework. Systems theory explains how diverse components and subsystems (such as employees and departments) benefit the whole when they work together toward a common objective and generate synergy (von Bertalanffy, 1972). Relationships between the most common theories in the literature review and systems theory are summarized in the following paragraph.

Transformational leadership theory identifies the leadership style that can result in significant improvement of both individual and system-wide performance (Williams, 2003). Organizational benefits arising from synergy within the system occur only when leaders support diversity, according to the learning-and-effectiveness paradigm (Thomas & Ely, 1996). Synergy can also be developed between the internal subsystems and the external environment of an organization, according to stakeholder theory (Freeman, 1984). The human perceptions of such subsystems were explored by social identity theory (Tajfel, 1974). Turning the synergy of diverse groups into effective decisions, such as hiring practices, requires effective leadership and management, as addressed in decision-making models and in contingency theory (Fiedler, 1964). Lastly, employees should not only have appropriate qualifications, but also be compatible with the values of the organizational system as a whole, as proposed in the person-organization fit model (Bowen et al., 1991) and the extended fit model (Powell, 1998). The view of organizations as systems provides a conceptual framework that connects the diverse theories and constructs of this literature review.

Review of the Literature

This study explores female technology managers' perceptions of diversity hiring. The academic literature summarized below includes seminal and recent works related to these constructs. A seminal work is defined as a composition that "strongly influences later developments" (Room & Brewer, 2009). The academic literature regarding the study's constructs is then addressed separately. Since the studied population is female managers in technology companies, many of the reviewed studies involved gender diversity in the technology field. The other relevant literature is broad in scope and is reviewed herein thematically, separated by construct.

Transformational Leadership

The following works have advanced the body of knowledge related to leadership. Early leadership theories are addressed first, as they laid the groundwork for developing transformational leadership theory (Bass, 1985a). Brief explanations of these early leadership theories and the work by Bass and his colleagues are presented here for reference, in chronological order.

Early Leadership Theories

Contingency theory helped explain how organizational leadership style should be adapted based on factors within and outside of the organizational system (Fiedler, 1964). Contingency theory shows quantitatively that there is no single leadership approach that would be ideal for all situations (Fiedler, 1964). This is a basis for understanding transformational leadership, a theory that would be developed two decades later (Avolio & Bass, 1995).

Between the theoretical advancements of contingency theory (Fiedler, 1965) and transformational leadership theory (Bass, 1985a), Tajfel (1974) contributed to the study of teamwork, Vroom and Jago (1974) advanced the study of organizational decision making, and Freeman (1984) published his groundbreaking work on stakeholder theory. Tajfel's work on social identity theory explains that individuals are influenced by their affiliations with groups and how those groups are perceived (Tajfel, 1974). Social identity can impact employees' and leaders' perceptions of diversity in the workplace, as organizations are comprised of diverse groups and subgroups (Tajfel, 1974).

The normative and descriptive models of decision making developed by Vroom and Jago (1974) can help leaders determine when to make decisions independently and when to employ group decision making. The efficacy of their normative model supported the perspective that leaders are not always capable of making the best decisions unilaterally (Vroom & Jago, 1974). According to Vroom and Jago (1974), collective decision making yields superior results in certain situations. This is consistent with systems theory's principle of synergy, where diverse inputs generate an outcome that could not have been created by an individual component (von Bertalanffy, 1972).

The powers of synergy and group decision making are also consistent with stakeholder theory, which posits that corporations create the most value when they address the needs of all stakeholders, not only investors (Freeman, 1984). This stakeholder approach suggests that diversity of interests should be considered in organizational decision making, including the interests of employees, managers, and external stakeholders. Employers might consider competing stakeholder interests (Freeman, 1984) when developing and implementing hiring practices, since systems benefit from diverse inputs, components, and subsystems.

Dodd (1932) provided an early example of the stakeholder concept, identifying investors, employees, consumers, and the public as groups with importance to corporations. The first known use of the term stakeholder in a business context occurred in a 1963 Stanford Research Institute memorandum (Donaldson & Preston, 1995; Freeman et al., 2010). Theorists including Ansoff, Abrams, Cyert, and March debated the responsibilities of corporations to stakeholders at this time (Freeman et al., 2010). Taylor, arguing in favor of corporate responsibility to stakeholders, predicted in 1971 that investors would eventually have decreased importance for corporations (Freeman et al., 2010). Freeman et al. (2010) distributed a working paper on stakeholder management, which resulted in the development of a seminar for AT&T managers (Freeman et al., 2010). Freeman's collaborations with AT&T and sociologist Evan led to the refinement of stakeholder theory (Freeman et al., 2010). The theory was introduced with the 1984 publication of Strategic Management: A Stakeholder Approach (Berman & Johnson-Cramer, 2017).

The basis of stakeholder theory is that businesses are responsible to create value for all stakeholder groups equally, not only investors (Donaldson & Preston, 1995). These stakeholder groups include a company's workforce, consumers, local communities, vendors, and investors (Donaldson & Preston, 1995). Examples of stakeholder value creation include career potential for workers, mutually beneficial relationships with vendors, fairly priced products for consumers, and fiduciary treatment of shareholder investments (Harrison & Wicks, 2013; Freeman & Dmytriyev, 2017). This emphasis on the direct impacts of the organization allows managers to consider and address the needs of key stakeholder groups (Freeman, 1984). Rather than balancing competing interests, the view of stakeholder theory is that value created for one stakeholder results in value creation for the others as well (Parmar et al., 2010). Stakeholder theory rejects the view that businesses must decide between maximizing shareholder wealth and the interests of other stakeholder groups (Berman & Johnson-Cramer, 2017).

Today's highly competitive business environment presents challenges that organizations have difficulty solving without the input of a diverse set of stakeholders (Harrison & Wicks, 2013; Freeman et al., 2010). Stakeholder theory recognizes the Kantian principle that people are more than means to ends (Freeman et al., 2010). Involving diverse stakeholders in organizational decision making not only provides valuable perspectives that benefit the company, it allows those with a stake in the outcomes to contribute to the deliberations (Freeman et al., 2010). Macaulay et al. (2018) explained that stakeholder theory promotes diversity even at the board of directors level.

Some companies began to track social performance and the awareness of social issues in business grew contemporaneously with stakeholder theory (Freeman et al., 2010). Milton Friedman famously claimed that corporations do not have a responsibility to society and should focus only on profit maximization (Donaldson & Preston, 1995; Freeman & Dmytriyev, 2017). Stakeholder theory, to the contrary, asserts that a business can simultaneously increase value to all stakeholder groups (Harrison & Wicks, 2013; Donaldson & Preston, 1995). Instead of choosing between satisfying investors or benefiting society, a corporation can address the needs of all direct stakeholders, thereby acting in both its financial and societal interests. Freeman's revolutionary work on stakeholder theory was quickly followed by groundbreaking research on transformational leadership from Bass (1985a) and his contemporaries.

Transformational Leadership Theory

Transformational leadership is defined as the act of motivating an employee to accomplish more than they had previously planned to accomplish (Bass, 1985a). Transformational leadership was addressed by Bass and Stogdill in 1981, but further developed over the following four years. The Avolio and Bass full range leadership model posits that the transformational leadership style can be more effective at improving organizational performance than the transactional and laissez-faire styles of leadership (Borgmann et al., 2016). As a result, transformational leadership can increase the level and sustainability of both employee performance and organizational performance (Jiang et al., 2017). The theory was endorsed, though critiqued, by Yukl (1999), who argued that it failed to explain the processes and underlying behaviors leaders use to transform organizations. For example, the literature does not specify how transformational leaders perceive diversity or how these perceptions influence hiring practices.

Bass (1985a) explained that transformational leadership was first contrasted with transactional leadership by James McGregor Burns. Bass developed transformational leadership theory to explain further how this style differs from transactional and laissez-faire leadership styles and how it can benefit contemporary organizations. Businesses must continually adapt to environmental changes, so it is often necessary to employ transformational leaders to ensure that the necessary adaptations occur (Bass, 1985a). If a company or department needs to maintain the status quo for a period of time, transactional leadership is appropriate; on the other hand, companies that need to adapt can benefit from transformational leadership (Bass, 1985a).

Understanding the difference and how to shift between these approaches is therefore relevant for

organizations and leaders who want to compete in today's dynamic global business environment.

The essence of transformational leadership is that those in power must know what type of

leadership they are expected to display (Bass, 1985a), again showing the influence of Fiedler's

contingency theory (1964). A leader should adapt their approach based on whether transactional

leadership or transformational leadership is needed by the organization; the former involves

achieving tactical objectives whereas the latter requires strategic thinking and employee

inspiration (Bass, 1985a). The transformational leader is able to inspire employees to accomplish

tasks they otherwise would not have completed or believed possible to achieve (Bass, 1985a).

Transformational leadership shares the principles of stakeholder theory, suggesting that

an organization should seek to address the needs of all who have a stake in it, including

employees, customers, suppliers, creditors, regulators, and local communities (Bass &

Steidlmeier, 1999). Transformational leaders use their power to advance the interests of all

stakeholder groups. They inspire others, are committed to learning, and oversee major

organizational changes (Bass & Steidlmeier, 1999).

Transformational leadership theory was falsely criticized as being supportive of immoral

leaders, such as political dictators (Bass & Steidlmeier, 1999). Bass and Steidlmeier (1999) drew

on extensive research to demonstrate that transformational leadership has a moral component and

does not apply to charismatic leaders who wield their power in an unethical fashion. These

theorists successfully defended transformational leadership and explained that its criticisms

should be directed to the practice of pseudo-transformational leadership (Bass & Steidlmeier,

1999). That is, leaders who appear to be transformational, yet who personally pursue a personal agenda, should not be viewed as transformational leaders. The moral failings of such leaders prevent them from attaining the status and accomplishments of a truly transformational leader.

According to Bass and Steidlmeier (1999), Fairholm explained in 1991 that diversity is a value intrinsic to transformational leadership. "Transformational leadership is value-centered. Leader and followers share visions and values, mutual trust and respect, and unity in diversity" (Bass & Steidlmeier, 1999, p.202). This brings us to the literature on diversity hiring, the study's second construct.

Diversity Hiring

This study's review of the literature on diversity hiring is divided into four parts. The first part is general research on organizational diversity. The second part is perceptions of diversity. The third part is group diversity, focusing on teams as subsystems of organizations. The final part is diversity hiring practices.

Organizational Diversity

Organizational diversity is variation in gender, ethnicity, generation, relation to organization, seniority, power level, and work expertise (Harjoto et al., 2015), as well as the multiplicity of views, knowledge, and approaches to work offered by members of varied identity groups (Thomas & Ely, 1996). Organizational diversity initiatives have increased in recent decades, aiming to improve organizational performance (Thomas & Ely, 1996), as indicated in the literature.

While organizational diversity could theoretically increase synergy and organizational performance, the research literature has not always found this true. Efforts have been made to understand the dynamics of organizational diversity better. The learning-and-effectiveness paradigm suggested that diversity only leads to performance improvement when it is supported by leaders (Thomas & Ely, 1996). Attempts have been made to quantify perceptions of diversity (Montei et al., 1996) while the justification–suppression model sought to explain how feelings about diversity are expressed (Crandall & Eshleman, 2003). The diversity impact navigator model aimed to help managers understand how to launch diversity programs successfully through simple and financially justified implementations (Wondrak & Segert, 2015). Yet the complex dynamics of organizational diversity have yet to be fully understood.

Perceptions of Diversity

Perceptions of diversity and their influence within organizational systems have been explored in the literature. Research has demonstrated that perceptions can be strongly influenced by the critical views of others (Bougheas et al., 2015) and that a leader's diversity perceptions can be influenced by their prior experience with a worker (Daniels et al., 2017). The organizational diversity-learning framework was recently developed as a method for addressing the challenges that can arise from diversity perceptions; the framework prepares employees to collaborate with colleagues regardless of perceived status, role, or background (Fujimoto & Härtel, 2017).

Despite efforts such as this to enhance organizational diversity, women and ethnic minorities who demonstrate appreciation for diversity often receive lower performance ratings as a result (Hekman et al., 2017). This is an aspect of the glass ceiling phenomenon that confronts women and ethnic minorities, who are prevented from moving into higher corporate positions, a topic Bass and Avolio addressed in 1994. They found that the glass ceiling limits upward mobility for women, though they are more likely than men to be transformational leaders (Bass & Avolio, 1994).

One thread of research has focused specifically on perceptions of gender diversity in the information technology (IT) field. Women were reportedly underrepresented in certain IT roles, although most study participants viewed diversity positively (Kohl & Prikladnicki, 2018). Perceptions of women's abilities to solve inequality were associated with women's attributions of responsibilities to do so (Kim et al., 2018). Positive perceptions of organizational diversity were found to influence female students, increasing their interest and comfort in the minimally diverse field of computer programming (Kulkarni et al., 2018). Role models, mentoring, and the engagement young girls in computing and STEM all contributed to an increase in female students selecting these fields of study (Bennaceur et al., 2018). Despite being underrepresented in certain IT roles, women were found to still value diversity (Kohl & Prikladnicki, 2018). In fact, identity conflict and feelings of exclusion exist among female computer scientists at all career levels, as a result of the lack of diversity in their field (Falkner et al., 2015).

Multiple factors have been identified that contribute to the lack of diversity in IT. Barriers to gender diversity in the recruitment of software development students resulted from unfamiliarity with computer science before college, societal views that males are better at computer science, and gender stereotypes regarding career opportunities in software development (Borsotti, 2018). Messages encouraging women to "lean in" to challenges sometimes shift the burden of achieving equality onto the shoulders of women (Kim et al., 2018). Also, diversity initiatives that fail to address systemic gender bias can make it difficult for women to both detect and address workplace discrimination, especially women who hold what are considered to be benevolent sexist beliefs (Brady et al., 2015) . It is evident from the literature that the perceptions of diversity are complex and not fully understood.

Group Diversity

Another thread of research examines diversity within groups, which are subsystems of organizations. An early study found that heightened diversity within a work unit was associated with diminished psychological attachment between coworkers within the unit (Tsui, Egan, & O'Reilly, 1992) , suggesting that diversity is a complex construct that does not automatically yield beneficial results. Studies have attempted to explain the impact of diversity on work groups, such as the dynamics of decision making in a diverse group.

Studies of group decision-making results are inconsistent (Kerr & Tindale, 2004) . As a social construct, group decision making in organizations is highly complex and can be influenced by many factors, not only diversity (Kerr & Tindale, 2004) . Ethnic and gender diversity are not as influential on team outcomes as is having diverse skill sets among the team members (Horwitz and Horwitz, 2007) .

Since there is not a general theory for group decision making (Hirokawa & Johnston, 1989) , various group decision-making methods have been presented (Chiravuri et al., 2011; Curseu et al., 2016) . One factor in effective group decision making is the preferred decision-making styles of the diverse members (Sager & Gastil, 1999). Group decision making is especially relevant for this book study because diversity hiring practices are established by organizational leadership.

More recently, there has been a significant thread of research related to corporate board gender diversity. As groups, corporate boards are central to achieving objectives, such as corporate social responsibility (CSR), yet few studies had measured the influence of board member diversity (Rao & Tilt, 2016). The presence of women on corporate boards was found to influence organizational financial performance, with CSR being a mediating variable (Galbreath, 2018). However, other aspects of board diversity have not been clearly linked to organizational outcomes (Kirsch, 2018). It is clear from this thematic review that the academic literature does not thoroughly explain the dynamic influences diversity can have on groups and organizations.

Diversity Hiring Practices

Diversity hiring practices are defined herein as the considerations and practices of diversity as a selection factor when hiring (Montei et al., 1996). Employees should share the norms and values of the organization that employs them, for the purpose of person-organization fit (Chatman, 1989; Bowen et al., 1991), except when there is justification for extending fit to include diversity (Powell, 1998).

Diversity management and inclusivity should inform hiring practices, as these efforts have a positive influence on the job satisfaction and performance of employees (Ohunakin et al., 2019). Diversity management in hiring even helps foster employee appreciation and innovation (Mazibuko & Govender, 2017). While employees of all backgrounds value diversity and diversity management, perceptions of diversity practices are significantly influenced by employee backgrounds (Kundu & Mor, 2017). What's more, a lack of inclusivity can lead to groupthink and prevent an organization from enjoying the benefits of diversity (Frost, 2018).

Studies of hiring practices have shown that leadership can influence perceptions of diversity and recruitment results. Both leadership and training were found to impact the acceptance of diversity within a police department's culture (Afful, 2018), for example.

Similarly, federal law enforcement agencies that do not actively recruit officers miss opportunities to increase the diversity of their applicant pools through targeted recruitment (Yu, 2018). The field of software engineering has also been shown to underperform in the recruitment of diverse job candidates, namely female college graduates (Menezes & Prikladnicki, 2018). Even high-achieving female applicants are often viewed negatively by recruiters, making it more difficult for them to obtain desirable jobs (Quadlin, 2018). Recruiters were found to favor male candidates who were qualified and committed, but only favored female candidates who were viewed as likeable (Quadlin, 2018). To address this sort of discrimination in hiring practices, diversity and inclusivity can be incorporated into the hiring strategies of organizations, promoting both employee satisfaction and performance (Ohunakin et al., 2019).

Synthesis of the Research Findings

The reviewed literature relates to transformational leadership and diversity hiring, especially the hiring of women and ethnic minorities. When leaders create a culture that values diversity, the varying perspectives and talents of the workforce can generate synergy, benefitting the organization (Thomas & Ely, 1996). Transformational leadership recognizes both the importance of organizational diversity and the necessity of unity (Bass & Steidlmeier, 1999). However, the broad range of findings in the literature suggests that each person's perceptions of diversity are unique and subjective.

A wide variety of factors have been found that influence perceptions of diversity (Bougheas et al., 2015; Daniels et al., 2017; Kulkarni et al., 2018). For example, various factors contribute to perceptions of gender diversity among female workers, students, and job applicants (Brady et al., 2015; Bennaceur et al., 2018; Borsotti, 2018). Hiring managers and recruiters'

28

perceptions of gender diversity have been shown to influence hiring practices (Hekman et al., 2017; Falkner et al., 2015; Kim et al., 2018; Kohl & Prikladnicki, 2018). In addition, models and instruments have been developed for the study and practice of diversity (Montei et al., 1996; Crandall & Eshleman, 2003; London et al., 2019; Fujimoto & Härtel, 2017).

The importance of hiring practices in shaping organizational diversity has been established (Quadlin, 2018; Afful, 2018; Yu, 2018; Menezes & Prikladnicki, 2018). Diversity's contribution to effective decision making and organizational outcomes has similarly been identified (Horwitz & Horwitz, 2007; Rao & Tilt, 2016; Galbreath, 2018; Kirsch, 2018). The value of diversity is well-established, but perceptions of diversity and methods to foster it are less clear.

Organizations and leaders can benefit by identifying the traits that act as barriers to equity (Wondrak & Segert, 2015; Kundu & Mor, 2017; Mazibuko & Govender, 2017; Frost, 2018; Ohunakin et al. 2019). Selecting and developing leaders with diversity-valuing traits can aid in this effort. However, managers resist innovation because of fear (Ackoff, 2006), which could contribute to reluctance to implement diversity initiatives. Some organizations embrace this change while others resist or struggle to adapt their hiring practices (Hekman et al., 2017; Quadlin, 2018). For this reason, this dissertation study explores the perceptions of female managers who practice transformational leadership and seek insights into diversity hiring practices within their technology companies.

The constructs of transformational leadership and diversity hiring have not been explored extensively in combination, according to a survey of the available literature. Most studies have used quantitative methodologies to measure diversity, its precedents, and its impacts, but have not explored the perceptions of diversity hiring among female managers in technology companies. Understanding how diversity is perceived by these leaders requires a qualitative approach, since that is the methodology used to study perceptions (Merriam & Tisdell, 2016). The research methods utilized in the relevant literature will now be summarized and critiqued.

Critique of Previous Research Methods

The majority of the seminal articles that were reviewed for this study have been theoretical, though some have had the quantitative support of primary or secondary data. The theories they propose are mostly based on the experiences and insights of the authors, as well as their reviews of existing literature. The recent research articles, on the other hand, are mostly quantitative studies.

The research articles are primarily objectivist, using quantitative analyses to test hypotheses. Data were gathered using both experimental and non-experimental methodologies, such as surveys and one audit study. These quantitative studies have varied levels of credibility, resulting from a wide spectrum of rigor in their methodologies. However, the findings from the quantitative studies were fairly consistent.

A few qualitative and mixed-methods studies have been found. These studies sought mostly to explore phenomena, such as the perceptions of diversity by women in the technology field. None of the studies explored the perceptions of diversity hiring by female managers in technology companies. The disproportionate number of quantitative studies suggests that there is likely an opportunity for more qualitative research. A conceptual framework has been developed for this research, through which to explore the interaction of the study's constructs.

Summary

This book study involves the constructs of transformational leadership and diversity hiring practices, among female managers in technology companies. It draws from various theories and is viewed through a conceptual framework based on systems theory. The relevant literature is therefore broad in scope and draws on multiple theories. The most relevant of these are transformational leadership theory, stakeholder theory, and the theory of person-organization fit.

The literature on leadership and diversity does not explore the perceptions of diversity hiring among female managers who are transformational leaders. The relevant research articles are mostly quantitative studies and are objectivist, using quantitative analyses to test their hypotheses. Data for these studies were gathered primarily from surveys. Despite this wealth of data, very few recent qualitative studies have been found that are relevant to the research questions. The qualitative methodology is therefore appropriate for this study because it deals with perceptions (Latham, 2016) .

Details regarding this study's methodology are provided in the following chapter. This includes explanations of the study's design, participants, and setting. The research questions are provided and the study's credibility and reliability are supported. In addition, methods for data collection and analysis are explained, along with the relevant ethical considerations.

CHAPTER 3. METHODOLOGY

This chapter presents the research methodology for studying the perceptions of diversity hiring among female managers in technology companies. The purpose of the study is to advance the body of knowledge related to transformational leadership and diversity hiring practices, particularly in technology fields. The chapter is divided into the research problem, research question, research design, target population and sample, procedures, instruments, and ethical considerations.

Purpose of the Study

Authors writing about transformational leadership style (Borgmann et al., 2016; Kirsch, 2018; Galbreath, 2018; Ohunakin et al., 2019) have not directly addressed diversity hiring. Research addressing both of these constructs directly was not found during the review of the literature. However, the literature has produced a broad background of research on which this study is based. This research is summarized in the following paragraphs, separated by construct.

The research literature on transformational leadership indicates that employees can be motivated to perform at levels higher than they expected or desired (Chai et al., 2017). Transformational leadership style can be more effective at improving organizational performance than the transactional and laissez-faire styles of leadership (Bass, 1985b). As a result, transformational leadership can increase the level and sustainability of employee job performance (Jiang et al., 2017). Transformational, transactional, and laissez-faire styles leadership styles were described in the Avolio and Bass full range leadership model (Borgmann et al., 2016).

The research literature on diversity indicates that diversity hiring contributes to organizations because of the variety of perspectives and approaches that it adds within organizations (Mazibuko & Govender, 2017). This includes the knowledge, abilities, and perspectives that previously underrepresented populations can contribute within their workplaces (Meric et al., 2015). As a result, effective diversity hiring has been found to enhance innovation, market access, employee retention, and organizational image (Wondrak & Segert, 2015).

The research literature on diversity hiring practices also indicates that a favorable perception of diversity leads to a positive view of incorporating underrepresented groups (Nakui et al., 2011). Conversely, an unfavorable perception of diverse populations results in mistrust and avoidance of individuals from such groups (Quadlin, 2018). As a result, leaders can influence workforce diversity by modeling appreciation for its benefits and fostering inclusive organizational cultures (Afful, 2018).

The purpose of the study, therefore, is to explore the perceptions of female managers in relation to diversity hiring. This exploration will hopefully provide insights to help these managers optimize diversity hiring practices. These insights could benefit the managers directly. They could also provide benefits for organizational stakeholders and relevant academic literature. To achieve this aim, the following research questions were developed.

Research Questions

The research questions for this study are

- RQ1: Which personal attributes of employees should be diverse within an organization?

- RQ2: For which jobs should diversity be a criterion in employee selection?

- RQ3: In which organizational situations should diversity be a criterion in employee selection?

Answering the research questions has required an appropriate methodology and design.

Research Design

This study follows the qualitative methodology of generic qualitative inquiry. In qualitative research, the researcher serves as the instrument, so their biases should be identified and distinguished from the participant's input (Merriam & Tisdell, 2016). The goal is not to judge the participant's perspective, but to describe and hopefully understand their experience (Erlingsson & Brysiewicz, 2013).

This study's approach utilized coding of data from semi-structured interviews, to comprehend the perceptions of the participants. The objective of the study is not to comprehend the unadulterated essence of the construct (Dowling, 2007), but simply to understand the participants' perceptions. There are several qualitative methodologies available to identify the essence of a construct, such as the descriptive, hermeneutic, lifeworld, interpretative, and dialogal approaches (Finlay, 2012). Qualitative researchers, for example, can apply Husserl's method of intuition, Heidegger's emphasis on the human experience, or Giorgi's focus on psychology (Gill, 2014). However, the generic qualitative inquiry methodology was appropriate for this qualitative study (Percy et al., 2015).

Target Population and Sample

This section describes the population and sample used in selecting the study's participants. Since the study's objective is to explore the perceptions of female managers in technology companies, a general population had to be defined (Merriam & Tisdell, 2016).

From there, a sample was recruited to participate in the study.

Population

The population for this study is female managers with transformational leadership style,

who work in technology companies within the United States. A sampling frame was required so

an appropriate population sample could be invited and selected to participate in the study. The

sample size had to be large enough to obtain a sufficient amount of data for analysis (Latham,

2016).

Sample

The sample was drawn from LinkedIn.com and the International Leadership

Association. Twelve participants were selected, who self-identified as female transformational

leaders and who worked in management positions in technology companies within the United

States. Their job titles ranged from supervisor to CEO.

Procedures

The procedures for this study began with taking a purposive sample of the population,

so that data could then be collected and analyzed (Latham, 2016). The size of the sample could

not be too large, since qualitative research prioritizes the appropriateness (or quality) of the

participants over the size of the sample (O'Reilly & Parker, 2013; Cleary et al., 2014).

However, a sample size that is too small can result in a lack of depth and breadth. When there

is sufficient data for the study to be replicated, it is no longer possible to manage the data, or

new information is not being gleaned from additional participants, then the sample's size limit

has been reached and interviewing should be discontinued (Fusch & Ness, 2015; Cleary et al.,

35

2014). For this study, an analysis of the relevant literature was made to determine typical sample sizes for similar qualitative studies. The analysis of comparable studies was the basis for selecting this sample size.

Participant Selection

The sample included 12 female technology managers who are transformational leaders. LinkedIn.com and the International Leadership Association were chosen as sources from which to recruit participants, because their memberships include female managers in technology companies. An invitation to participate was posted on the association's website and sent directly to LinkedIn members. Interested parties were invited to apply for participation. Potential participants responded by email.

The screening process confirmed the title, gender, geographic location, and technology industry of each participant. This is criterion sampling, which involves the development of required criteria for the participants to meet (Cleary et al., 2014). These criteria included the participant's demographic attributes and their leadership experience. Criterion sampling is a purposive form of sampling as it targets participants ideally suited for the study (Merriam & Tisdell, 2016).

Specific protocols were followed for the selection of participants. Their privacy and protection were similarly ensured by strict procedures, as was data collection. These processes are detailed in the following sections.

The participants were currently employed as managers in technology companies. Participants were not relatives or colleagues of the researcher. They each self-identified as female and confirmed working in the United States. Each answered in the affirmative when asked if they perceived themselves as a transformational leader. All recruiting materials clearly stated that transformational leadership was part of the eligibility criteria for participation in the study. A $50 honorarium was offered to each participant who completed the study.

Protection of Participants

The risk that a participant's identity would be determined by their interview answers being included in the study was minimal. However, caution was taken to identify and omit any material that would reveal personally identifiable information. There were no other likely risks beyond those typically involved with participation in any study.

An informed consent form was provided to each participant via email before their interview. Each consent form was signed and returned before an interview was held. The content of the form was reviewed with the participant before their interview began, allowing them the opportunity to ask any questions they had before the interview.

Demographic data including current employer, gender identity, ethnicity, and race were collected. The purpose of collecting these data types was to ensure that the participants were female technology managers and to understand whether they were ethnic or racial minorities. Knowing their minority status is important in understanding whether the participants' perspectives represented a broad or narrow range of ethnic backgrounds. These data were kept confidential and have not been included in the study's findings.

Audio recordings of the interviews and other documentation are stored on an encrypted hard drive. Files have been labeled by participant number and participant names were not included in the documents or recordings. Analysis of electronic records was conducted at the location where the hard drive is stored, so it did not need to be transported.

Alpha-numeric codes were used to identify participants immediately and stored separately, per standard procedures for academic research (Saunders et al., 2014). Personal identifiers and consent forms have been stored only in hard copy. They are kept in a locked location at an educational institution. After seven years, paper documentation will be shredded, audio recordings will be permanently deleted from online storage, and the hard drive containing electronic records will be erased and then physically destroyed.

Expert Review

The interview questions were refined with the help of an expert review panel. The panel consisted of a researcher of leadership, a change management consultant, and a human resources director. Each panelist has extensive experience with transformational leadership, diversity, and hiring practices.

Meetings were arranged that included the panelists and the faculty mentor, through two video-conferencing sessions. One panelist participated in the first session. The other panelists participated on the following day, due to conflicting schedules.

During each session, the attendees were welcomed and introduced. This was followed by an explanation of the session's objective of refining the interview questions. Panelists were given the opportunity to ask questions or seek clarification.

The panel was then presented with each interview question one at a time and asked to provide feedback on how the question might be interpreted by the study's participants. Field notes were taken during the sessions and key points were repeated to the panelists, to check for accuracy. Constant comparison was made between the in vivo panelist discussion, research objectives, and field notes, for the purpose of triangulation (Patton, 2014).

The panelists were thanked for their participation. Based on feedback from the panelists, two of the initial questions were deemed to be redundant. The remaining questions were refined for clarity and are the initial guiding interview questions used in this study.

Data Collection

Data collection for this study involved triangulation and associated methods. The collection of data was in preparation for analysis of the data, which used open, axial, and selective coding (Williams & Moser, 2019). Triangulation was achieved through the use of interviews, observations, and artifact review (Patton, 2014). Triangulation of the interviews, observations, and field notes enhanced this study's credibility, transferability, dependability, and confirmability. These are the four elements of validity in qualitative research as identified by Guba and Lincoln (Trochim, 2006).

Semi-structured interviews were the primary sources of data, since they allow a researcher to obtain data from participants in a manner that offers structure, yet flexibility as needed (Latham, 2016). The semi-structured interviews utilized a consistent set of guiding questions, but the flow of each interview depended on the data that emerged and observations that were made of each participant. Follow-up questions were added as appropriate, for clarification or additional detail (Willgens et al., 2016).

Interview data were collected over video-conferencing software, due to the coronavirus pandemic. Listening intently to participants during interviews was very important to optimize the confirmability of the results. The audio from these 30-to-60-minute sessions was recorded and later transcribed (Saunders et al., 2014). Member checks were conducted after each interview, to confirm that the content and affect of each participant's experience were properly understood (Bower, 2017).

The data collected from the participants was very consistent and aligned with the study's theory, model, and rationale. By the twelfth interview, it was determined that there was sufficient data for the study to be replicated and new information was not being gleaned from additional participants. Per standard practices in qualitative studies, it was determined that the sample's size limit had been reached and interviewing was then discontinued (Fusch & Ness, 2015; Cleary et al., 2014).

The interview sessions also allowed each participant's physical setting, body language, gestures, tone of voice, and environment to be observed and noted. These observations occurred at the beginning and end of each session, but the participant video was turned off for the entire period in which interview questions were asked and answered. These observational notes augmented the interviews, in the form of memos (field notes). Such notes documented observations and reflections on both the data and the process (Salmona & Kaczynski, 2016), allowing the recording of thoughts in an unconstrained manner. The observations were compared with the participant responses, to triangulate data and link themes relative to the dissertation's focus (Patton, 2014).

Artifacts were another useful source of data for the study. Efforts were made to note or even obtain documents presented by the participants. No participant documents were received, however. Instead, relevant artifacts observed during the interviews were a source of data used for triangulation (Latham, 2016). The use of video-conferencing technology prevented the close inspection of artifacts. As a substitute, field notes were kept during interviews and observations.

Care was taken to prevent the field notes from interrupting the interview and observation processes. Artifact review did not reveal a significant amount of information. However, documenting artifacts that are visible, using memos and other notes to record how a researcher interacts with the study's process, enhances a study's credibility (Merriam & Tisdell, 2016).

Comparing interview data with researcher observations allowed for the collection of both content and affect data, in support of the theory of perception (Bower, 2017). This helped demonstrate that interpretations of the participant experiences had been confirmed by multiple sources of data. If they had not agreed, that would have suggested bias or misunderstanding of the participant experiences (Patton, 2014).

Data Analysis

To explore the perceptions of the participants, analysis of the dozens of pages of data that had been collected from the interviews, observations, and artifact review occurred through open, axial, and selective coding (Williams & Moser, 2019). Notes, observations, artifacts, and insights were documented while conducting the interviews and while reading the interview transcripts, consistent with common qualitative research practices (Merriam & Tisdell, 2016).

To prepare for analysis, the interviews and notes were entered into separate documents and a spreadsheet, a process which required transcription (Merriam & Tisdell, 2016). This stage was followed by member checks to confirm accuracy of intent and content, coding of the data, and further analysis (Merriam & Tisdell, 2016).

Open coding was used to define thematic fragments and concepts in which to assemble the data (Williams & Moser, 2019). This initial step allowed axial codes to be appropriately established, refining and organizing the thematic material into categories (Williams & Moser,

2019). The constant comparison method was employed throughout the analysis, so that codes could evolve through multiple rounds of comparison (Willgens, 2016). Selective coding involved the integration and prioritization of categories into themes (Williams & Moser, 2019). These emergent themes aligned with the research questions and the following matrix. Representative quotes from the interviews were then selected to help illustrate each theme. From this analytical framework, themes emerged (Kaefer et al., 2015), as indicated in the following table.

Table 1

Data Analysis Matrix

Research Questions	**Primary Codes**	**Categories**	**Themes**
Which personal attributes of employees should be diverse within an organization? (Powell, 1998)	race gender beliefs/opinions experience ethnicity personality	background attributes intellectual attributes personal attributes physical attributes professional attributes	Diversity is valued in relation to a wide variety of employee personal attributes.
For which jobs should diversity be a criterion in employee selection? (Powell, 1998)	board of directors vice president manager programming entry-level internship	leadership non-leadership	Employee diversity is valued for all jobs.
In which organizational situations should diversity be a criterion in employee selection? (Powell, 1998)	international operations industry growth company growth research/development raising capital startup restructuring problem solving lack of diversity	company growth company decline lack of diversity	Employee diversity is valued for all organizational situations.

This qualitative research not only included collecting and analyzing data, but also strove to reach synthesis in its analysis. Data analysis is a challenging part of the analytical framework, so a systematic approach was used for data management and categorization (Latham, 2016). Open coding was used to identify categories in the data. This was followed by axial coding. Codes were used to identify and categorize data, allowing for the emergence of both insights and patterns (Merriam & Tisdell, 2016). Next, a selective coding process was followed where the developed categories were tested and substantiated (Merriam & Tisdell, 2016). In this way, synthesis was achieved from the key themes that emerged from the data analysis.

The organization and analysis of the study's data used the following coding scheme. Transcripts of the interviews were prepared, accompanied by memos. Each transcript was read while notes and detailed observations were documented throughout, as recommended by Merriam and Tisdell (2016). Insights that arose from reflection on the interviews were also documented, beginning the process of pattern recognition, theme development, and general analysis (Merriam & Tisdell, 2016). Notes, observations, and insights were then reviewed carefully to analyze and synthesize the data further.

Identifying links between categories allowed for higher-order categories than what might have been accomplished otherwise (Bringer et al., 2006). Observational notes were also compared with the interview data. Comparing the data sources provides data triangulation and increases the credibility of results (Patton, 2014).

Instruments

This section addresses both the role of the researcher and the guiding interview questions. No survey instruments were used in the study since the data were collected through in-depth interviews over Zoom conferencing software. Validity and reliability of a study instrument are therefore not relevant for this study.

The Role of the Researcher

The role of the researcher is to collect and analyze data (von Alberti-Alhtaybat & Al-Htaybat, 2010). This requires attention to validity, dependability, design accuracy, ethics, honesty, and abiding by the legacies of research traditions (Merriam & Tisdell, 2016). Research methodologists advance the body of knowledge in their discipline by adding new information or assessing the accuracy of existing information. This has traditionally been accomplished in quantitative studies by presenting and testing a hypothesis (Latham, 2016). Qualitative research, on the other hand, can be used to document and analyze the experiences of individuals, providing insights into their perspectives. In quantitative analysis, the researcher strives to remove their biases from the study, while the qualitative researcher recognizes and documents their biases, prejudices, and influences on the study (Merriam & Tisdell, 2016).

Various approaches to qualitative research have been developed including phenomenology, ethnography, grounded theory, narrative inquiry, and case studies (Merriam & Tisdell, 2016). Each approach considers the role of the researcher based on the nature of the methodology and the intent of the study. Regardless of the approach, the researcher serves as the study's instrument, so identifying and documenting bias is paramount (von Alberti-Alhtaybat & Al-Htaybat, 2010). This is achieved through bracketing, which is an essential analytic methodology. Bracketing was conducted before the study began and was useful in separating researcher bias from participant data (Latham, 2016). In addition to bracketing, qualitative research methods were studied and applied to the interview process for this study.

Guiding Interview Questions

A consistent framework was followed for each of the interviews. The participant was welcomed and reminded of the purpose of the study. The informed consent document was reviewed and the participant was given an opportunity to ask questions. Demographic data were collected. They were informed that they could end the interview at any time. Confirmation of their permission for the use of an audio recording device was obtained. The one-hour anticipated length of time for the interview was provided. The participant was reminded that there was no requirement to answer questions that they did not want to answer.

The interviews were semi-structured. Participants were asked to respond to each of the statements. Follow-up questions were used as necessary to obtain clarification and elaboration from the participants regarding their perceptions of diversity, as transformational leaders.

The first three guiding interview questions were used for the earlier interviews. For compliance purposes, three additional interview questions were added for later interviews, under guidance provided by the study's committee chairperson.

1. In your view as a technology leader, what is the influence of transformational leadership style on your perceptions of diversity?

2. Similarly, in your view, what is the influence of these diversity perceptions on your hiring practices?

3. How does transformational leadership style influence your perceptions of diversity and hiring practices?

4. In your experience as a transformational leader hiring for person-organization fit, which employee personal attributes warrant including diversity as a criterion?

5. In your experience as a transformational leader hiring for person-organization fit, which jobs warrant including diversity as a criterion?

6. In your experience as a transformational leader hiring for person-organization fit, what organizational situations warrant including diversity as a criterion?

Ethical Considerations

This study protects the rights of its participants under procedures approved by Capella University's Institutional Review Board. The population is not identified as being vulnerable. The participants are female managers in technology companies. While participating in viable and productive research, the study made every reasonable effort to avoid any degree of harm to participants.

Qualitative research is not free from ethical considerations (Latham, 2016). This study ensured that participants understood the nature of the study and their participation. Confidentiality has been maintained as outlined in the study's approved plan. The participants were not placed in danger of any kind. Given the in-depth nature of some participant interviews, extra effort was taken to ensure that participants did not experience mental or emotional trauma as a result of their involvement in the study. Ultimate responsibility to uphold these ethical principles rests with the researcher (Merriam & Tisdell, 2016).

The research followed the ethical standards of the Academy of Management. This includes disclosing conflicts of interest, assumptions, theories, methods, and metrics (Academy of Management, n.d.). In addition, the assumptions of cited research and findings have been reported fully and accurately (Academy of Management, n.d.).

Qualitative research involves the following steps: selecting a topic, developing a research problem, establishing a theoretical framework, reviewing the relevant literature, selecting a sample, conducting the study, and reporting the findings (Merriam & Tisdell, 2016). There are no conflicts of interest that have arisen from this study (Academy of Management, n.d.). The development of the research problem, theoretical framework, and research methodology considered any risks and ethical assumptions that were most likely to arise during the study (Merriam & Tisdell, 2016).

The study's literature review section accurately and thoroughly reports the findings, methods, and assumptions of relevant research (Academy of Management, n.d.). The report of the study's methodology identifies how participants were recruited and selected. The participants were each offered 50 dollars as compensation for participating in the study. Care was taken to inform participants of the study's purpose and risks before obtaining their full consent (Merriam & Tisdell, 2016). Interviews were conducted in a manner that minimized risk and discomfort for the participants, especially since qualitative studies ask participants to describe their personal experiences with the relevant constructs. Qualitative studies can involve interview questions that are personal and reflective (Percy et al., 2015), causing participants to feel discomfort or even embarrassment. Sensitivity to participant feelings and thoughts allowed participant discomfort to be detected and the sessions redirected accordingly, as necessary.

Systems are in place to safely store and keep confidential all private or personally identifiable participant data. This allows participant anonymity and has required procedures for removing participant names, locations, and backgrounds from publication and even from the unnecessary view (Saunders et al., 2014). This required secure information management systems and processes, preventing the data from being viewed by those without permissions. The sampling and recruiting methods have been identified and the study's assumptions, limitations, and findings have been transparently reported, in conformity with Academy of Management standards (Academy of Management, n.d.).

Each participant has provided their informed consent to participate in the study. Their audio recordings and the transcripts of the interviews have been kept confidential. An anonymization system has been used to label the data so that the participant names, employers, and locations were not stored with the data (Saunders et al., 2014).

Summary

Chapter 3 identified the study's research problem, research questions, and exploratory qualitative design. A qualitative methodology was chosen to explore the perceptions of diversity hiring, among female managers who are transformational leaders in technology companies. The target population and sampling procedures were outlined, providing justification for the selection of the final participants. The participants have all met the study's inclusion criteria and been recruited through either LinkedIn.com or the International Leadership Association.

Sound procedures to protect the participants and collect data were outlined, specifying the detailed processes that were followed to ensure data accuracy and privacy (Saunders et al., 2014). The constant comparison method was utilized in conjunction with coding and researcher notes, for the purposes of triangulation (Patton 2014). Survey instruments were not used in the study. The role of the researcher and the interview process have been explained, along with an assessment of the study's ethical considerations. A presentation of the data and analysis is provided in Chapter 4.

CHAPTER 4. PRESENTATION OF THE DATA

Introduction: The Study and the Researcher

This chapter provides a description of the study's sample, the research methodology that was used to collect and analyze the data, and a presentation of the findings that resulted from the analysis. Due to the qualitative nature of the study, the researcher's role is to serve as the study's instrument (von Alberti-Alhtaybat & Al-Htaybat, 2010). This has been accomplished through a careful process of analysis, to find themes and patterns arising from the data and relevant to the research questions (Williams & Moser, 2019). Care has been taken to identify and ignore the biases of the research, allowing the data to be examined and analyzed with no anticipated outcome (Erlingsson & Brysiewicz, 2013). Detailed discussion of the findings will be provided in the next chapter.

Description of the Sample

The sample included twelve female managers in the technology field. Each participant worked in the United States and identified as a transformational leader. The participants were interviewed to ascertain their perceptions and gain insights regarding the study's research questions. Three additional applicants for the study, who were initially approved to participate, did not schedule interviews, for personal reasons unrelated to the study.

The study participants were diverse in terms of their employers, ages, geographic locations, ethnic backgrounds, and job titles. The diverse nature of the sample enhances transferability of the study's findings to other scenarios (Merriam & Tisdell, 2016). When there is a high degree of variation within a sample, there will be a larger population with similarities to the participants (Merriam & Tisdell, 2016).

Protection of the Participants

An informed consent form was provided to each participant in advance of their interview. The participants each signed their consent forms and were reminded of their rights before their interviews commenced. No questions or objections to the terms of the consent form were raised by the participants.

Research Methodology Applied to the Data Collection and Data Analysis

This generic qualitative study is exploratory in nature. Data collection included in-depth semi-structured interview questions, observations, and artifact review. Triangulation was achieved through the comparison of the data that had been collected through these methods (Patton, 2014). Analysis was conducted using open, axial, and selective coding (Williams & Moser, 2019).

Data Collection

The interviews were limited by the use of distance technology, due to the coronavirus pandemic. Semi-structured interview questions were used as guides in each interview. Follow-up questions were added as appropriate, for clarification or additional detail (Willgens et al., 2016). Listening intently to participants during interviews was very important to optimize the confirmability of the results. An interview transcript was provided to each participant, providing the opportunity to correct any errors in transcription. Accuracy of interview transcriptions increases reliability and validity, based on how closely the findings of the study correlate with the constructions of reality that the participants have created (Merriam & Tisdell, 2016).

The interviews were conducted over Zoom video-conferencing software. Observations of relevant data were made during each participant's Zoom session. These observations were compared with the participant responses, for the purpose of triangulating data and linking themes relative to the focus of the book (Patton, 2014).

Artifact review, while an important element of data collection, could not be fully realized. Participants did not provide any documents. Further, the use of video-conferencing technology prevented the close inspection of artifacts. As a substitute, field notes were kept during interviews and observations. Care was taken to prevent the field notes from interrupting the interview and observation processes. Artifact review did not reveal a significant amount of information. However, documenting a researcher's position using memos and other notes, to record how the researcher interacts with the study's process, enhances a study's credibility (Merriam & Tisdell, 2016) .

Triangulation of the interviews, observations, and field notes enhanced this study's credibility, transferability, dependability, and confirmability. These are the four elements of validity in qualitative research as identified by Guba and Lincoln (Trochim, 2006) . Credibility is established through the confidence with which the findings of the study can be implemented, based on the quality of the data collection and analysis (Latham, 2016) . Transferability is the ability to apply the findings in other contexts (Trochim, 2006) . Dependability in qualitative research is analogous to reliability in quantitative research; dependability involves strict adherence to the research methodology accounting for changes to the research context, such as maintaining an audit trail of the data collection and analysis processes (Guba & Lincoln, 2001) to maintain consistency between the data and the results (Merriam & Tisdell, 2016) . Confirmability is the degree to which the study's findings can be replicated by others (Trochim, 2006) . These four elements of research validity were firmly established, resulting from the careful triangulation of the interviews, observations, and artifact review.

Data Analysis

The analysis incorporated the dozens of pages of data collected from the interviews, observations, and artifact review. Three types of coding were employed: open, axial, and selective. Open coding was used initially, to define thematic fragments and concepts in which to assemble the data (Williams & Moser, 2019). The following steps were establishing axial codes and then refining and organizing the thematic material into categories (Williams & Moser, 2019). Codes evolved through multiple rounds of comparison, through use of the constant comparison method throughout the analysis (Willgens, 2016). The integration and prioritization of categories into themes was achieved through selective coding (Williams & Moser, 2019). These emergent themes aligned with the research questions and Table 1. This matrix contains codes, categories, and themes on one axis; the other axis is Powell's (1998) three justifications for including employee diversity as an aspect of organizational fit. Representative quotes from the interviews were then selected to help illustrate each theme.

Presentation of Data and Results of the Analysis

The analysis of the data aimed to answer the research questions, related to the perceptions of diversity hiring, among female managers who are transformational leaders in technology companies. Specifically, the research questions address the perceptions of including diversity as a criterion in the employee selection process. This inclusion of diversity is referred to as extended fit (Powell, 1998). Including diversity as a criterion places value on employee characteristics that do not fit the organization's existing employee norms, but instead complement them.

52

The study's results are provided below and address the three research questions:

- RQ1: Which personal attributes of employees should be diverse within an organization?

- RQ2: For which jobs should diversity be a criterion in employee selection?

- RQ3: In which organizational situations should diversity be a criterion in employee selection?

Research Question 1

Research Question 1 asked, "Which personal attributes of employees should be diverse within an organization?" The purpose is to explore which attributes should be diverse, from their experiences. The participants universally mentioned race and gender as beneficial types of diverse personal attributes among employees. In addition to race and gender, the participants collectively identified 28 other personal attributes that they felt should be diverse among employees. These personal attributes can be broadly categorized as: background, intellectual, personal, physical, and professional. Each category will now be addressed separately.

Background Attributes

The background attributes identified by participants included employee personal experience, culture, national origin, upbringing, and privilege. Having employees with diversity in these areas was identified as being beneficial. Following are relevant statements from the interviews.

Participant 8 valued diverse employee personal experiences and stated, "I have tried really hard to introduce…experiential diversity." Participant 12 indicated that the technology field was disproportionately comprised of European American men, but that it has evolved to include more employees who are from India or Asia, addressing the diversity of culture and national origin. Participant 3 raised the issue of privilege and mentioned hiring employees who "may not have had the advantages that others may have had."

Intellectual Attributes

Diverse intellectual attributes among the employees can also be beneficial, according to the participants. The participants identified the following attributes, which have been classified herein as intellectual: beliefs, perspectives, priorities, communication styles, education levels, languages, religions, and social skills.

Participant 3 provided the following perspective:

I don't believe an organization can transform itself with all the same voices, all the same humans who…don't have any differences. And so I would struggle to think of an example of an organization that has succeeded in transforming without having diversity of thought, diversity of opinion.

Participant 7 specified that differences in social skills and communication styles should be valued. "If...their social skills...the way they communicate is...really, really different...if I'm managing somebody who's different from me...it would be up to me to preserve what they have and not have them just become everyone else." Similarly, Participant 2 values curiosity among employees and indicated that curious workers expect to see diversity among team members. "I bring a lot of young team members to the team because they are curious. They want to learn; they want to understand and see some diverse team members."

Personal Attributes

Diversity among personal attributes can also be beneficial, according to the participants.

The category of personal attributes includes geographic location, personality, alma mater, marital status, parental status, and style of dress. Employee diversity regarding these personal attributes is a sign of a healthy organizational culture, as evidenced by the following participant comments.

Participant 5 advocates for hiring people with "nontraditional backgrounds that may not meet criteria with coming from particular universities…" When screening job applicants, Participant 3 will "filter for willingness to learn from others, filter for willingness to be wrong, filter for willingness to say, 'I don't know, but I'll find out.'" Participant 10 took pride in her organization having "different countries that are coming together to work together to put something out for the projects that we're working on." Participant 11 shared an example of a pregnant job candidate whom her boss discouraged the participant from hiring:

> I am good with someone being pregnant with their third child. And I was really insulted that he even brought that notion up, that I'm about to hire a pregnant woman...A pregnant woman and a man who's having a child, in the view of an employer, should be the same because nowadays men do take paternity leave as well...It was a judgment in judging her, because she had gotten pregnant out of wedlock.

Physical Attributes

Race and gender were the most frequently mentioned types of diversity. Regarding gender, participants universally specified gender as an employee attribute that should be diverse, but did not specify biological sex. In addition to advocating for diversity of race and gender, there are other physical attributes that should be diverse, according to the participants. These include ethnicity, ability, age, LGBT (lesbian, gay, bisexual, transgender) status, and body type.

Participant 3 defined diversity in this way: "Diversity is...different genders, different gender expressions, different races, different ethnicities, different abilities, different ways of looking at the world, cultures...the gamut of human experience, and having as much of that as you can." Participant 12 shared a similar sentiment:

I want diversity of thought, I want diverse people, I want diverse age groups, you name it, I want it as long as they have the skill set. And I do put that as a priority in my hiring practices.

Participant 7 provided an insight into a unique ability that female and non-white leaders have in modeling racial and gender diversity.

I've been an example of somebody who's not your...stereotypical Asian or even more so...Asian female...I think it's a form of representation that's probably important so that...it creates future diversity in leadership as well. And it changes the perception of the mainstream about what's possible.

Professional Attributes

Diversity of employees in relation to professional attributes was also valued by the participants. These professional attributes included which industries the employees had worked in, prior job levels, professional credentials, and length of service with the organization. The following excerpts illustrate these preferences.

Participant 3 valued diversity at all levels of the organization:

Having folks who are literally on the ground talking to customers and executives who are C-suite folks... folks who have been doing it for years and years, and folks who are brand new to it and maybe coming in from a new experience... from a different industry or a different point of view, folks who have had different perceptions, folks who may

need/want/desire a bigger focus on benefits over salary or vice versa.

Similarly, Participant 4 values diverse professional backgrounds among employees:

When I say diversity, it's not just how you look and feel. Right? It's also the diverse

experiences that have brought you to the table for typically our industry…For me,

because this function is so derived within health information technology and dependent

on the overarching health care ecosystem, I wanted to diversify the talent pool also on

where we're pulling this talent from.

In summary, the analysis of the data showed that there is a broad range of diverse

employee personal attributes that the participants valued. The diversity of employee race and

gender were universally recognized as being beneficial for organizations. However, diversity of

background, intellectual attributes, personal attributes, physical attributes, and professional

attributes were also viewed as beneficial. As Participant 12 summarized:

If we're trying to solve for something in a way that's not ever been done before, we need

as many diverse views, thoughts, experiences as possible on the team. If everybody's got

groupthink, we all think alike. We all have similar experiences. We're not going to get

very far and so it totally informs how I hire.

Research Question 2

Research Question 2 asked, "For which jobs should diversity be a criterion in employee

selection?" The participants identified jobs that warrant including diversity as a criterion. These

jobs ranged from the bottom to the top of the organizational hierarchy. At the highest level,

diversity was valued among members of the board directors. According to Participant 6, "The board of directors was all older white men and I pushed and pushed and pushed to get a woman on the board." Diversity among executives and organizational leaders was valued as well. Participant 6 also identified a common practice by executives within technology companies as hiring "their male colleagues into the leadership roles rather than looking from within or reaching out to communities that represent more diverse people."

The participants suggested that technology companies benefit from hiring transformational leaders who value diversity and who are themselves diverse. Participant 6, for example, explained that she serves as a role model to other diverse employees: "Being a representative of a minority in tech, a woman myself, is important because it's really hard to imagine being something that you cannot see."

This emphasis on the role of transformational leadership in diverse organizations raised the prospect of whether transformational leaders can thereby promote diversity through all of the four factors from transformational leadership's model. These factors are commonly referred to as the "Four Is" and include intellectual stimulation, idealized influence, individual consideration, and inspiring motivation (Hall et al., 1969). While participants self-identified as transformational leaders and responded to questions based on their own perspectives relative to transformational leadership style, following are excerpts from the participant interviews, which appear to align with the factors of the Four I model.

Factor 1: Intellectual Stimulation

This leadership factor can foster diversity by building a diverse, representative, and collaborative group with intellectual humility, curiosity, and the courage to innovate, according to the participants.

Participant 12 shared this analogy:

Captain Kirk is a transformational leader. He's the one that's looking…beyond the stars,

beyond the heavens and…using his wit and asking for input from Spock...If we're trying

to solve for something in a way that's not ever been done before, we need as many

diverse views, thoughts, experiences as possible on the team.

Participant 9 emphasized that "leaders who are in those positions of power, they really do

set the culture, the tone, even the make-up of who is in the team."

Factor 2: Idealized Influence

This leadership factor can foster diversity by creating human connections and

recognizing those characteristics that make each person unique, according to the participants.

Participant 1 offered an anecdote of a transformational leader who had demonstrated

idealized influence:

He would ask you about your dog or your kid or whatever, and the fact that he was able

to make these human connections and really connect with this team...the team was

willing to jump in and carry the load for him, just because of the respect and admiration

that they had for him.

Participant 9 summarized idealized influence in this way: "Ultimately it comes down

to...how do we treat each other as human beings? And then you move away from intellectual

stimulation to relationship-based interactions in the workplace."

Factor 3: Individual Consideration

This leadership factor can foster diversity by viewing each individual holistically and providing the resources, support, and opportunities they need to continually improve, according to the participants.

Participant 7 described the objective of individual consideration: "How do I bring out the best of however you are in whatever form that is?"

Participant 9 provided the following summary of individual consideration. "We as a culture need to start looking at the whole human being...inclusive of their health, mental, emotional, physical, spiritual, intellectual stimulation...that all plays a lot into someone's fulfillment in the workplace."

Factor 4: Inspirational Motivation

This leadership factor can foster diversity by promoting an inclusive and cohesive team with a clear sense of purpose, according to the participants.

Participant 12 provided the following insight into inspirational motivation. "I set vision and I create what I always call a north star...The way that transformational leadership can integrate diversity into the ecosystem is to...not just talk about it, but integrate it in a way that's compelling."

This was expounded upon by Participant 5:

Having that leadership style helps promote diversity in hiring practices because there's a comfortability with challenging your own biases and...hiring folks who are not like you and who may not conform to your preconceived notions of who may qualify for a job and may not conform with some preconceived notions around what the ideal candidate looks like for the job.

While diversity in senior leadership roles was highly valued, jobs at lower levels also benefit from employee diversity, according to the participants. These positions included middle management, program managers, product managers, project managers, entry-level positions, and even interns. An effort to increase diversity among entry-level candidates was provided as an example: "We're partnering now where we're training sophomores in college to learn our technology so that they can get jobs here upon graduation."

Participant 2 made it clear that there were no jobs where diversity was not a criterion in hiring. "I want to look at all jobs for diversity." In contrast with Powell's suggestion that jobs with decision-making authority require warrant higher levels of employee diversity, these transformational leaders find value in diversity even for entry-level positions.

Research Question 3

Research Question 3 asked, "In which organizational situations should diversity be a criterion in employee selection?" The organizational situations that warrant including diversity were varied, much like the aforementioned personal attributes and jobs. From a corporate strategy perspective, these situations included times of industry growth, business formation, and company growth. Participant 2 stated that industry growth requires that the employees reflect the nation as a whole: "I need more people that are America's base, which means very diverse." Participant 7 described the value of diverse perspectives among executives at start-ups, which she did not appreciate until later in her career: "We're very like go-getter, idealist…and our VP of business operations was very conservative and the one to pick out what might go wrong and…at the time, I didn't appreciate it at all."

Participants also identified solving problems and innovating, as situations that warrant including diversity as a hiring criterion. For example, Participant 7 provided this anecdote to illustrate how her diverse perspective cut through groupthink and prompted deeper reflection: "I'd sit in on these review meetings with designers…The design team that was especially touchy feely. And I'd be like, so why'd you put the button over there? Silence across the room."

Other organizational situations that warranted including diversity as a criterion were connecting with customers and conducting international business. Participant 5 related that "transformational leadership style…is more readily open to diverse opinions…helping you connect with the consumer audience…to get the implications of the data and understanding of the data." Participant 10, whose employer serves foreign customers, highlighted the importance of cultural and linguistic diversity when conducting business internationally. "Most of the people that are on the projects that I'm working on, they come from different…areas in the world."

Another situation that warrants prioritizing diversity in hiring is when barriers to diversity exist in the organization. The barriers to diversity identified by the participants fell into five categories: managerial bias, lack of diversity among qualified applicants, recruiters not being committed to diversity, transactional leadership, and a lack of organizational inclusivity.

Specific examples of managerial bias were those that favored particular races, genders, ages, religions, political views, alma maters, and whatever employee characteristics a "narcissistic" leader might find self-serving. Participant 1 felt that managerial bias was strongest in mid-level managerial positions. "I think where the gap is, is in our middle management."

Participant 7 made a point that removing bias reduces discrimination: "It's not about hiring out more gay people, but at least not discriminating on the fact that they are."

Participants attributed the lack of diversity among qualified applicants primarily to a lack of representation of diverse groups within the organization. Participant 10 shared her personal experience with this:

> Most of the companies that I work with as a technical person...less than one percent of the people in my field were women...The programming area that I chose was more of a field that had to do with...being up till...4:00 a.m., working on the hardware side of things...it was very much...a male figure would be out there doing the things that I was doing. So I did find myself being sometimes the only female in that area.

Other factors that reduce diversity among job applicants include the geographic locations of some jobs and the qualifications that some women feel they lack. Participant 6 reported, "Unless women meet all of the qualifications, they tend to not apply."

Recruiters not being committed to diversity is a barrier to diversity that was associated primarily with a lack of creativity in selecting recruitment sources. The participants suggested that hiring practices that promote diversity involve reducing bias and making diversity a priority throughout the recruiting process, as well as using targeted sources to reach a wider spectrum of candidates. Participant 5 recommended specific practices:

> Job descriptions are written in a way that they would attract a lot of different candidate types...Be mindful of your own biases...challenging the recruiting team to look for nontraditional backgrounds that may not meet criteria with coming from particular universities or education or experience credentials.

To increase diversity among applicant pools, Participant 1 shared the following recruiting practices:

> We've partnered...with nonprofits, organizations that are focused on those areas in which we don't have the people...Hispanic Association of Colleges and Universities...Girls Who Code...opportunity on-ramps in which students go to an internship six months and then they get a job...We've done returnship, with people that have left the workforce and are now coming into the workforce. We are training military.

Transactional leadership (in contrast with transformational leadership) is another barrier to diversity. This leadership style sometimes results in a lack of inclusivity in organizations, remote work that can feel alienating, and a focus on short-term results. Participant 2 drew a connection between leadership style and appreciation of diversity:

> I'm looking more now more than ever for curious minds. So when I'm doing interviews, I'm asking certain questions to understand whether their minds are curious and are looking forward to learning more...I notice more and more that the diverse team members are more in line with my transformational thinking than mainstream team members.

The lack of organizational inclusivity was described as a white-male-dominated hierarchy and culture, especially by the participants from information technology firms. Participant 6 estimated, "Women-run businesses get somewhere around two to three percent of all venture capital funding...So we're leaving...dollars on the table because...venture capitalists aren't being proactive and looking to fund these women-run companies or underrepresented underestimated-led companies." To diversify technology organizations, Participant 3 recommended looking beyond tech stereotypes to find the ideal candidates:

Filter for willingness to learn from others; filter for willingness to be wrong; filter for willingness to say, 'I don't know, but I'll find out.' Those are all things that I hire for. I hire for curious; I hire for…relentlessness, but at the same time, balancing that with healthy boundaries and…health first, family second, work a very distant third.

The situations that warrant including diversity as a hiring criterion are broad, according to these participants and transformational leaders. These situations can be categorized as periods of company growth, periods of company decline, and when barriers to diversity exist.

Summary

The research questions related to technology managers' perceptions of diversity in hiring. The questions asked about the personal attributes, jobs, and organizational situations that the participants believe warrant including diversity as a hiring criterion. The findings presented above suggest that diversity is itself a diverse concept. By this is meant that the participants value diversity among a wide variety of employee personal attributes, not only racial and gender diversity.

The jobs that warrant including diversity as a hiring criterion are likewise diverse. They span from intern to board member, including all levels in between. Diversity among organizational leaders is especially important as this provides the added benefits of attracting applicants and inspiring employees, according to the participants.

Following suit, the organizational situations that warrant prioritizing diversity in hiring are similarly broad, as described by the participants. Whether an organization is just beginning or an industry leader and whether it is focused on analyzing consumer data or developing products, diversity provides benefits, according to these transformational leaders. However, barriers to diversity can take many forms and be deeply entrenched in organizational and individual biases. Overcoming these barriers requires managers and recruiters to prioritize diversity in all stages of the recruitment and hiring process. Managers then need to maintain an inclusive culture where all employees feel that their unique characteristics are valued, in furtherance of the collective organizational vision. These findings will be discussed in detail in the following chapter.

CHAPTER 5. DISCUSSION, IMPLICATIONS, RECOMMENDATIONS

This chapter includes discussion, implications, and recommendations arising from the data analysis that was presented in chapter 4. This will include a summary, a discussion, and the conclusions of the results. Limitations, implications, and recommendations will be included as well.

Summary of the Results

The objective of the study is to advance the body of knowledge related to transformational leadership and diversity hiring practices, particularly in technology companies, by exploring female managers' perceptions of diversity hiring. Specifically, the study's overall aim is to explore the perceptions of these managers regarding including diversity as a criterion in the employee selection process.

This was accomplished through in-depth interviews of female managers with transformational leadership style, employed by technology companies. The participants provided insights into their perceptions of diversity hiring. Including diversity as a criterion places value on employee characteristics that do not fit the organization's existing employee norms, but instead complement them, according to the participants. The inclusion of diversity as a criterion has been explained by Powell (1998) as an extension of organizational fit.

The conceptual framework for the study allowed for relationships between multiple theories to be recognized, in relation to the constructs of leadership and diversity hiring. Systems theory provides the foundational view that organizations are open systems (von Bertalanffy, 1972). These systems benefit from the diversity of their subsystems and components, which are the organizational stakeholders (Freeman, 1984). Transformational leadership promotes the harmonious interaction of the stakeholder groups (Bass & Steidlmeier, 1999). Organizational leaders should include diversity as a hiring criterion in order to avoid focusing too much on organizational fit, which can lead to groupthink and related issues (Powell, 1998).

The study identified personal attributes, jobs, and organizational situations for which diversity should be a hiring criterion, according to the participants. Participant statements were provided for context, to illustrate participant perspectives. In short, the findings suggest that diversity is itself a diverse concept. The findings also reveal that participants value diversity among a wide variety of employee personal attributes, jobs, and organizational situations, as the following paragraphs indicate.

The participants all mentioned race and gender as beneficial types of diverse personal employee attributes. In addition to race and gender, the participants collectively identified 28 other personal attributes that should be diverse among employees. These personal attributes have been categorized for this study as: background, intellectual, personal, physical, and professional.

The jobs where employee diversity is valued by the participants range from intern to board member, including all levels in between. Diversity among organizational leaders is especially important as this provides the added benefits of attracting applicants and inspiring employees. This does not minimize the importance of diversity among employees who are not leaders, however. Even diversity among entry-level employees and interns was valued.

Organizational situations where employee diversity is valued are similarly broad. Whether an organization is growing or in decline, beginning or established, raising capital or engaged in research and development, diversity provides benefits, according to the participants. An organization that lacks diversity is another situation when diversity should be prioritized, eliminating organizational and individual barriers to diversity. This requires managers and recruiters to prioritize diversity in all stages of recruitment and hiring. Managers then need to maintain an inclusive culture so all employees feel that their unique characteristics are valued.

Discussion of the Results

The study addresses three research questions. The findings that arose in relation to each research question are now provided. The primary themes that arose from the data analysis are provided as well, in relation to each research question.

Research Question 1

Research Question 1 asked, "Which personal attributes of employees should be diverse within an organization?" The primary theme that arose from RQ1 was that diversity is valued in relation to a wide variety of employee personal attributes. The diversity of employee race and gender were universally recognized as being beneficial for organizations. However, diversity of background, intellectual attributes, personal attributes, physical attributes, and professional attributes were also viewed as beneficial. Participant 12 provided the following statement, which illustrates that diversity is valuable among a broad array of personal attributes.

If we're trying to solve for something in a way that's not ever been done before, we need as many diverse views, thoughts, experiences as possible on the team. If everybody's got groupthink, we all think alike. We all have similar experiences. We're not going to get very far and so it totally informs how I hire.

Research Question 2

Research Question 2 asked, "For which jobs should diversity be a criterion in employee selection?" The primary theme that arose from RQ2 was that employee diversity is valued for all jobs. These jobs ranged from the bottom to the top of the organizational hierarchy. At the highest level, diversity was valued among members of the board directors and executive team. While diversity was highly valued in board and executive roles, jobs at lower levels also benefit from employee diversity. These positions included middle management, program managers, product managers, project managers, entry-level positions, and even internships.

Participant 5 suggested that the best candidates for any job might have characteristics that differ from what is expected. For this reason, diversity hiring can help identify the candidates who are best suited for open jobs. In the words of Participant 5:

Having that [transformational] leadership style helps promote diversity in hiring practices because there's a comfortability with challenging your own biases and...hiring folks who are not like you and who may not conform to your preconceived notions of who may qualify for a job and may not conform with some preconceived notions around what the ideal candidate looks like for the job.

Research Question 3

Research Question 3 asked, "In which organizational situations should diversity be a criterion in employee selection?" The primary theme that arose from RQ3 was that employee diversity is valued for all organizational situations. These situations included times of industry growth, business formation, and company growth. Participants value diversity during problem solving, innovation, market research, and international operations. Special importance was placed on diversity when there is a lack of diversity within the organization, resulting from either individual or systemic barriers. These barriers to diversity fell into five categories: managerial bias, lack of diversity among qualified applicants, recruiters not being committed to diversity, transactional leadership, and a lack of organizational inclusivity. To diversify technology organizations, specific diversity hiring practices were recommended.

Participant 1 shared the following recruiting practices that her organization had utilized, in various situations and stages of organizational growth:

> We've partnered...with non-profits, organizations that are focused on those areas in which we don't have the people...Hispanic Association of Colleges and Universities...Girls Who Code...opportunity on-ramps in which students go to an internship six months and then they get a job...We've done returnship, with people that have left the workforce and are now coming into the workforce. We are training military.

There was no particular outcome that had been anticipated for this study. It is noteworthy that the participants spoke with a unified voice in calling for diversity of all types, in all jobs, and in all organizational situations. It is also of interest that every applicant for the study exhibited transformational leadership style, even those who were unfamiliar with transformational leadership theory. These findings emerged from the data collection and analysis, suggesting that the methodology for the study was appropriate and effective.

Conclusions Based on the Results

The perceptions of the participants suggest that they value diversity among a wide variety of employee personal attributes, jobs, and organizational situations. The participants identified 28 personal attributes that should be diverse among employees, in addition to race and gender. These personal attributes have been categorized for this study as: background, intellectual, personal, physical, and professional.

The jobs where employee diversity is valued by the participants range from intern to board member. Special importance was given to diversity among organizational leaders since this diversity aids in attracting applicants and inspiring employees. The importance of diversity among employees who are not leaders should not be disregarded, according to the participants. Diversity was valued even among entry-level employees and interns.

The participants did not identify any occasions when diversity would not be a beneficial hiring criterion. Diversity benefits organizations that are growing or in decline, new or established, raising capital or engaged in research and development, according to the participants. Organizations that currently lack diversity should strive to overcome organizational and individual barriers to diversity. This requires that the hiring personnel prioritize diversity in all stages of the recruitment and selection process. Managers should maintain an inclusive culture where all employees feel valued for their unique characteristics and contributions.

The participants universally supported that the concept of organizational fit can be extended to include diversity. An organization whose workforce is too homogeneous runs the risk of falling into groupthink, therefore employee diversity extends organizational fit to maintain a healthy ecosystem of ideas and perspectives.

Comparison of Findings With Theoretical

Framework and Previous Literature

The findings are consistent with the study's conceptual framework. Systems theory (von Bertalanffy, 1957) is the overarching theory and has been applied to organizations, since they function as open systems. Such systems have diverse subsystems and components, which collaborate toward a common objective. The diversity of the components within an open system is analogous to the diversity of the employees. To foster the synergy that can occur within an organization system, transformational leaders apply the Four Is (Hall et al., 1969) to support employees as they pursue a common organizational vision.

Organizational systems require that the employees have the required characteristics to fit the organization, but also contribute to a diverse pool of perspectives. Hiring for organizational fit (Bowen at al., 1991) emphasized the former and gained tremendous popularity in the 1990s. However, when Chatman (1989) proposed organizational fit, she indicated that there is an optimal balance between the heterogeneity and homogeneity of the personnel. Powell (1998) sought to reconcile this dichotomy between heterogeneity and homogeneity by suggesting that organizational fit can be extended, to allow for an appropriate level of diversity. He concluded that diversity should be a hiring criterion for some employee personal attributes, jobs, and organizational situations. However, this study of female technology managers found that they value diversity for all employee personal attributes, all jobs, and all organizational situations.

Interpretation of the Findings

The findings of this study suggest that there tends to be insufficient employee diversity in technology companies. Increased diversity is recommended by the female technology managers who participated in the study. This increased diversity should not be limited to race and gender, but should include the many other personal attributes that comprise employee diversity. The increase should also not be limited to particular jobs or organizational situations. Instead, technology companies should prioritize diversity in all jobs and organizational situations,

73

especially when barriers to diversity exist. If an organization removes all barriers to diversity, a truly diverse workforce could hypothetically emerge, but no examples of such an organization were provided by the participants. This reinforces the conclusion that increased diversity is warranted in technology companies. It also suggests that organizational fit should be extended to include diversity, so that organizations do not become too insular and homogenous.

Organizations should include diversity as a hiring criterion for all positions. This accomplishes more than just meeting arbitrary DEI (diversity, equity, and inclusion) objectives. Prioritizing diversity in hiring might allow an organization to be more adaptable to the rapid changed of modern business, because of the multiplicity of perspectives and ideas that exist among its ranks.

Limitations

Limitations of the study are that it did not produce generalizable results or statistical correlations, due to the subjectivist nature of qualitative research (Latham, 2016). The study's results are limited to the insights that were generated from the input of the participants, all of whom are female managers in technology companies. Males, non-managers, transactional leaders, and employees of non-technical companies were not included in the study. Another limitation of the study is the researcher's personal biases, though every reasonable attempt was made to prevent bias from influencing the outcome of the study. Because of its qualitative nature, no causal relationships or definitive conclusions should be drawn from the findings of this study.

Implications for Practice

The findings suggest that hiring practices which promote diversity involve reducing bias, making diversity a priority throughout the recruiting process, and using targeted sources to reach a wider spectrum of candidates. Promoting transformational leadership within organizations could also foster diversity, since Fairholm found that diversity is a value intrinsic to transformational leadership (Bass & Steidlmeier, 1999). From their unique vantage point, female technology leaders appear to foster diversity and inclusiveness, benefiting their organizations as well as themselves. This contributes to the literature on diversity hiring and organizational fit, which previously suggested extending fit only in certain situations.

Diversity hiring practices can help foster diversity and inclusiveness. Nine diversity hiring practices recommended by the participants follow. Supporting statements from the participants are also included.

- Recruit from colleges and universities that serve minoritized and veteran populations. Participant 11 stated:

When I do get applicants from different backgrounds, I look at all of them…Some of your best people might have come from a Cal State University because they couldn't afford to go to a better university…I do make a conscious decision of looking at everything as a plus, not really an obstacle.

- Recruit through personal and professional networks. Participant 6 identified the following practices:

I look to groups who are communities who have a community of minorities in tech or diverse groups…use our networks, lift other people up…reaching out to communities that represent more diverse people…Go out to this local women-in-tech meetup that's happening.

- Write job descriptions that attract diverse candidates. Participant 5 suggested the following:

Job descriptions are written in a way that they would attract a lot of different candidate types…Be mindful of your own biases…challenging the recruiting team to look for nontraditional backgrounds that may not meet criteria with coming from particular universities or education or experience credentials.

- Post job announcements in places that serve minoritized and veteran populations. Participant 8 shared the following example:

When we see a candidate pool of all old men, we try to push each other and push our recruiters to open up different avenues and to kind of widen the applicant pool and make sure that we're getting as wide of an applicant pool and as wide of a diversity of candidates as we can.

- Partner with organizations that serve minoritized and veteran populations. Participant 1 stated:

We've partnered…with non-profits, organizations that are focused on those areas in which we don't have the people…Hispanic Association of Colleges and Universities…Girls Who Code…opportunity on-ramps in which students go to an internship six months and then they get a job…We are training military.

- Participate in local meetups for women, minoritized, and veteran populations. Participant 6 shared the following approaches:

Go out to this local women-in-tech meetup that's happening...introduce myself to..some

of the members of that community and just talk to them...I've seen people get leadership

roles out of Twitter. I myself got my latest job from being part of a [diversity] group...and

putting my profile on their website.

- Offer returnships (for those re-entering the workforce) and internships. Participant 1

  again recommended:

Opportunity on-ramps in which students go to an internship six months and then they get

a job... We've done returnship, with people that have left the workforce and are now

coming into the workforce. We are training military. We're partnering now where we're

training sophomores in college to learn our technology.

- Make job postings, applications, onboarding, and training accessible for those with

  disabilities. Participant 3 suggested "making...training accessible, making [the

  company's] marketing message accessible."

- Use anonymous resume screening to remove recruiter bias. Participant 3 advocated:

...Taking names off of resumes, taking...cities and states off of resumes, taking phone

area codes off of resumes so that you can go in as blind as possible... Filter for

willingness to learn from others, filter for willingness to be wrong, filter for willingness

to say, "I don't know, but I'll find out." Those are all things that I hire for.

These nine practices were recommended by the participants to help increased workforce

diversity. Increasing workforce diversity could result in a greater variety of perspectives and

talents within organizations (Chang et al., 2019; Mazibuko & Govender, 2017). As a result,

diversity can help improve organizational performance (Ohunakin et al., 2019). Specific benefits that diversity could provide are enhanced innovation, market access, employee retention, and organizational image (Wondrak & Segert, 2015). For these reasons, technology companies might benefit from extending fit to include diversity of all types, in all jobs, and in all organizational situations.

Recommendations for Further Research

Since this exploratory qualitative study could not produce generalizable results or statistical correlation, further studies should be conducted. Such studies could expand the qualitative research or quantify particular variables. The following diversity hiring research questions are proposed, to further the inquiry that was initiated by this study.

Why is employee diversity typically thought of in terms of race and gender, instead of other personal attributes? How does employee diversity change within an organization as it develops, from its founding to its peak and then into its stage of decline? How can barriers to diversity be effectively identified and addressed? Which hiring practices yield the best results for increasing employee diversity within organizations?

The following research questions related to leadership are also worth consideration for additional study. What is the ratio of transformational leadership style to transformational leadership style, among female technology managers? What is this ratio among male technology managers? Do organizations that are not technology companies have similar ratios of transformational leadership style to transformational leadership style, both for female and male managers? What are the differences in the perceptions of diversity hiring between female managers in information technology, biotechnology, and other technology companies?

78

Conclusion

This qualitative study has explored the perceptions of diversity hiring among female managers with transformational leadership style, in technology companies. The participants shared their experiences and perspectives, providing clear answers to the research questions. These managers value employee diversity in all forms, jobs, and organizational situations. An additional finding that emerged from the data is that every applicant demonstrated a strong preference for transformational leadership style, suggesting compatibility between this leadership style and diversity hiring practices.

Race and gender were identified as beneficial types of diverse personal employee attributes. However, 28 other personal attributes were identified that should be diverse among employees. These personal attributes were divided into the following five categories for this study: background, intellectual, personal, physical, and professional. This variety of personal attribute categories suggests that race and gender are not the only personal attributes that should be diverse within an organization.

Employee diversity was valued by the participants at all organizational levels. These levels range from intern to board member. Organizational leaders can attract applicants and inspire employees through their diversity. This does not negate the importance of diversity at other levels of the organization, including entry-level employees and interns.

Organizational situations where employee diversity is valued include growth and decline, startup and expansion, capital acquisition and product development, according to the participants. When organizational diversity is lacking, that is a situation when barriers to diversity should be overcome, both at the individual and organizational levels. This requires diversity to be prioritized in all stages of the recruitment and hiring process. Diversity should then be maintained by managers, through the creation of an inclusive culture where all employees feel valued for their unique characteristics.

It is hoped that this research provides insights to help technology companies and their members understand how organizational diversity is perceived and can be fostered for the benefit of all stakeholders. This will hopefully lead to decisions that increase diversity within organizations, resulting in greater organizational performance and outcomes for their stakeholder groups. Ideally, these findings might be particularly useful to female managers, transformational leaders, and researchers interested in diversity hiring and leadership.

REFERENCES

Academy of Management. (n.d.). *AOM code of ethics.* https://aom.org/about-aom/governance/ethics/code-of-ethics

Ackoff, R. L. (2006). Why few organizations adopt systems thinking. *Systems Research and Behavioral Science, 23*(5), 705-708. https://doi.org/10.1002/sres.791

Afful, I. (2018). The impact of values, bias, culture and leadership on BME under-representation in the police service. *International Journal of Emergency Services, 7*(1), 32-59. https://doi.org/10.1108/IJES-05-2017-0028

Aikin, S. F. (2019). Deep disagreement, the dark enlightenment, and the rhetoric of the red pill. *Journal of Applied Philosophy, 36*(3), 420-435. https://doi.org/10.1111/japp.12331

Allyn, B. (2020, December 17). Google AI team demands ousted Black researcher be rehired and promoted. NPR Technology. http://www.npr.org/2020/12/17/947413170/google-ai-team-demands-ousted-black-researcher-be-rehired-and-promoted

Arif, S., & Akram, A. (2018). Transformational leadership and organizational performance. *Seisense Journal of Management, 1*, 59-75. https://doi.org/10.5281/ZENODO.1306335

Avolio, B. J., & Bass, B. M. (1995). Individual consideration viewed at multiple levels of analysis: A multi-level framework for examining the diffusion of transformational leadership. *The Leadership Quarterly, 6*(2), 199-218. https://doi.org/10.1016/1048-9843(95)90035-7

Avolio, B. J., & Bass, B. M. (2001). B.J. Avolio & B.M. Bass (Eds.), *Developing potential across a full range of leadership TM: Cases on transactional and transformational leadership.* Taylor and Francis. https://doi.org/10.4324/9781410603975

Banks, G. C., McCauley, K. D., Gardner, W. L., & Guler, C. E. (2016). A meta-analytic review of authentic and transformational leadership: A test for redundancy. *The Leadership Quarterly, 27*(4), 634-652. https://doi.org/10.1016/j.leaqua.2016.02.006

Bass, B. M. (1985a). Leadership: Good, better, best. *Organizational Dynamics, 13*(3), 26-40. https://doi.org/10.1016/0090-2616(85)90028-2

Bass, B. M. (1985b). *Leadership and performance beyond expectations.* Collier Macmillan.

Bass, B. M., & Avolio, B. J. (1994). Shatter the glass ceiling: Women may make better managers. *Human Resource Management, 33*(4), 549-560. https://doi.org/10.1002/hrm.3930330405

Bass, B. M., & Steidlmeier, P. (1999). Ethics, character, and authentic transformational leadership behavior. *The Leadership Quarterly, 10*(2), 181-217. https://doi.org/10.1016/S1048-9843(99)00016-8

Bass, B. M., & Stogdill, R. M. (1981). *Stogdill's handbook of leadership: A survey of theory and research.* Free Press.

Bedi, A., Alpaslan, C. M., & Green, S. (2016). A meta-analytic review of ethical leadership outcomes and moderators. *Journal of Business Ethics, 139*(3), 517-536. https://doi.org/10.1007/s10551-015-2625-1

Bennaceur, A., Cano, A., Georgieva, L., Kiran, M., Salama, M., & Yadav, P. (2018). Issues in gender diversity and equality in the UK. *Proceedings of the 2018 ACM/IEEE 1st International Workshop on Gender Equality in Software Engineering,* 5-9. https://doi.org/10.1145/3195570.3195571

Birks, M., Chapman, Y., & Francis, K. (2008). Memoing in qualitative research: Probing data and processes. *Journal of Research in Nursing, 13*(1), 68–75. https://doi.org/10.1177/1744987107081254

Bogan, V.L., Just, D.R., & Dev, C.S. (2013). Team gender diversity and investment decision-making behavior. *Review of Behavioral Finance, 5*(2), 134-152. https://doi.org/10.1108/RBF-04-2012-0003

Borgmann, L., Rowold, J., & Bormann, K. C. (2016). Integrating leadership research: A meta-analytical test of Yukl's meta-categories of leadership. *Personnel Review, 45*(6), 1340-1366. https://doi.org/10.1108/PR-07-2014-0145

Borsotti, V. (2018). Barriers to gender diversity in software development education: Actionable insights from a Danish case study. *Proceedings of the ACM/IEEE 40th International Conference on Software Engineering: Software Engineering Education and Training,* 146-152. https://doi.org/10.1145/3183377.3183390

Bougheas, S., Nieboer, J., & Sefton, M. (2015). Risk taking and information aggregation in groups. *Journal of Economic Psychology, 51,* 34-47. https://doi.org/10.1016/j.joep.2015.08.001

Bowen, D. E., Ledford, G. E., & Nathan, B. R. (1991). Hiring for the organization, not the job. *Academy of Management Perspectives, 5*(4), 35-51. https://doi.org/10.5465/ame.1991.4274747

Bower, M. (2017). Husserl on perception: A nonrepresentationalism that nearly was: Husserl on perception. *European Journal of Philosophy, 25*(4), 1768-1790. https://doi.org/10.1111/ejop.12261

Brady, L. M., Kaiser, C. R., Major, B., & Kirby, T. A. (2015). It's fair for us: Diversity structures cause women to legitimize discrimination. *Journal of Experimental Social Psychology, 57*, 100-110. https://doi.org/10.1016/j.jesp.2014.11.010

Bringer, J. D., Johnston, L. H., & Brackenridge, C. H. (2006). Using computer-assisted qualitative data analysis software to develop a grounded theory project. *Field Methods, 18*(3), 245–266. https://doi.org/10.1177/1525822X06287602

Chai, D. S., Hwang, S. J., & Joo, B. (2017). Transformational leadership and organizational commitment in teams: The mediating roles of shared vision and team-goal commitment. *Performance Improvement Quarterly, 30*(2), 137-158. https://doi.org/10.1002/piq.21244

Chang, E. H., Milkman, K. L., Chugh, D., & Akinola, M. (2019). Diversity thresholds: How social norms, visibility, and scrutiny relate to group composition. *Academy of Management Journal, 62*(1), 144-171. https://doi.org/10.5465/amj.2017.0440

Chatman, J. A. (1989). Improving interactional organizational research: A model of person-organization fit. *The Academy of Management Review, 14*(3), 333-349. https://doi.org/10.5465/amr.1989.4279063

Chiravuri, A., Nazareth, D., & Ramamurthy, K. (2011). Cognitive conflict and consensus generation in virtual teams during knowledge capture: Comparative effectiveness of techniques. *Journal of Management Information Systems, 28*(1), 311-350. https://doi.org/10.2753/MIS0742-1222280110

Cleary, M., Horsfall, J., & Hayter, M. (2014). Data collection and sampling in qualitative research: Does size matter? *Journal of Advanced Nursing, 70*(3), 473–475. https://doi.org/10.1111/jan.12163

Conyon, M. J., & He, L. (2017). Firm performance and boardroom gender diversity: A quantile regression approach. *Journal of Business Research, 79*, 198-211. https://doi.org/10.1016/j.jbusres.2017.02.006

Crandall, C. S., & Eshleman, A. (2003). A justification-suppression model of the expression and experience of prejudice. *Psychological Bulletin, 129*(3), 414-446. https://doi.org/10.1037/0033-2909.129.3.414

Curseu, P. L., Schruijer, S. G. L., & Fodor, O. C. (2016). Decision rules, escalation of commitment and sensitivity to framing in group decision-making. *Management Decision, 54*(7), 1649-1668. https://doi.org/10.1108/MD-06-2015-0253

Daniels, D. P., Neale, M. A., & Greer, L. L. (2017). Spillover bias in diversity judgment. *Organizational Behavior and Human Decision Processes, 139*, 92-105. https://doi.org/10.1016/j.obhdp.2016.12.005

Dappa, K., Bhatti, F., & Aljarah, A. (2019). A study on the effect of transformational leadership on job satisfaction: The role of gender, perceived organizational politics and perceived organizational commitment. *Management Science Letters, 9*(6), 823-834. https://doi.org/10.5267/j.msl.2019.3.006

Dodd, E. M. (1932). For whom are corporate managers trustees? *Harvard Law Review, 45*(7), 1145-1163. https://doi.org/10.2307/1331697

Donaldson, T., & Preston, L. E. (1995). The stakeholder theory of the corporation: Concepts, evidence, and implications. *Academy of Management Review, 20*(1), 65-91. https://doi.org/10.2307/258887

Dowling, M. (2007). From Husserl to van Manen: A review of different phenomenological approaches. *International Journal of Nursing Studies, 44*(1), 131–142. https://doi.org/10.1016/j.ijnurstu.2005.11.026

DuBow, W. M., & Kaminsky, A. (2019). How an online women in technology group provides a locus of opposition. *Computers in Human Behavior, 98*, 285-293. https://doi.org/10.1016/j.chb.2019.05.006

Eberly, M. B., Bluhm, D. J., Guarana, C., Avolio, B. J., & Hannah, S. T. (2017). Staying after the storm: How transformational leadership relates to follower turnover intentions in extreme contexts. *Journal of Vocational Behavior, 102*, 72-85. https://doi.org/10.1016/j.jvb.2017.07.004

Erlingsson, C., & Brysiewicz, P. (2013). Orientation among multiple truths: An introduction to qualitative research. *African Journal of Emergency Medicine, 3*(2), 92-99. https://doi.org/10.1016/j.afjem.2012.04.005

Falkner, K., Szabo, C., Michell, D., Szorenyi, A., & Thyer, S. (2015). Gender gap in academia: Perceptions of female computer science academics. *Proceedings of the 2015 ACM Conference on Innovation and Technology in Computer Science Education*, 111-116. https://doi.org/10.1145/2729094.2742595

Fiedler, F. E. (1964). A contingency model of leadership effectiveness. *Advances in Experimental Social Psychology, 1*, 149-190. https://doi.org/10.1016/S0065-2601(08)60051-9

Finlay, L. (2012). Unfolding the phenomenological research process: Iterative stages of "seeing afresh." *Journal of Humanistic Psychology, 53*(2), 172–201. https://doi.org/10.1177/0022167812453877

Forrester, J. W. (2016). Learning through system dynamics as preparation for the 21st century. *System Dynamics Review, 32*(3-4), 187-203. https://doi.org/10.1002/sdr.1571

Freeman, R. E. (1984). *Strategic Management: A Stakeholder Approach.* Pitman Publishing, Inc.

Freeman, R. E., & Dmytriyev, S. (2017). Corporate social responsibility and stakeholder theory: Learning from each other. *Symphonya, 2017*(1), 7-15. https://doi.org/10.4468/2017.1.02freeman.dmytriyev

Freeman, R. E., Harrison, J. S., Wicks, A. C., Parmar, B. L., & De Colle, S. (2010). *Stakeholder theory. The state of the art.* Cambridge University Press.

Frost, S. (2018). How diversity (that is included) can fuel innovation and engagement – and how sameness can be lethal. *Strategic HR Review, 17*(3), 119-125. https://doi.org/10.1108/SHR-03-2018-0020

Fujimoto, Y., & Härtel, C. E. J. (2017). Organizational diversity learning framework: Going beyond diversity training programs. *Personnel Review, 46*(6), 1120-1141. https://doi.org/10.1108/PR-09-2015-0254

Fusch, P. I., & Ness, L. R. (2015). Are we there yet? Data saturation in qualitative research. *The Qualitative Report, 20*(9), 1408–1416. https://doi.org/10.46743/2160-3715/2015.2281

Galbreath, J. (2018). Is Board Gender Diversity Linked to Financial Performance? The Mediating Mechanism of CSR. *Business & Society, 57*(5), 863–889. https://doi.org/10.1177/0007650316647967

Galvan, J. L., & Galvan, M. C. (2017). *Writing literature reviews: A guide for students of the social and behavioral sciences* (7th ed.). Routledge.

Gardner, D. M., & Ryan, A. M. (2020). What's in it for you? Demographics and self-interest perceptions in diversity promotion. *Journal of Applied Psychology, 105*(9), 1062-1072. https://doi.org/10.1037/apl0000478

Gill, M. J. (2014). The possibilities of phenomenology for organizational research. *Organizational Research Methods, 17*(2), 118–137. https://doi.org/10.1177/1094428113518348

Gleibs, I. H., & Haslam, S. A. (2016). Do we want a fighter? The influence of group status and the stability of intergroup relations on leader prototypicality and endorsement. *The Leadership Quarterly, 27*(4), 557-573. https://doi.org/10.1016/j.leaqua.2015.12.001

Grant, N. (2022, Jun 13). Google agrees to give $118 million in relief to settle class-action pay discrimination case. *New York Times.* https://www.nytimes.com/2022/06/12/business/google-discrimination-settlement-women.html

Guba, E. G., & Lincoln, Y. S. (2001). Competing paradigms in qualitative research. In C. F. Conrad, J.G. Haworth, & L.R. Lattuca. (Eds.), *Qualitative research in higher education: Expanding perspectives* (2nd ed., pp. 57–72). Pearson Custom Publishing.

Hall, J., Johnson, S., Wysocki, A., & Kepner, K. (1969). Transformational leadership: The transformation of managers and associates: HR020/HR020, 7/2002. *EDIS, 2002*(2) https://doi.org/10.32473/edis-hr020-2002

Harjoto, M., Laksmana, I., & Lee, R. (2015). Board diversity and corporate social responsibility. *Journal of Business Ethics, 132*(4), 641-660. https://doi.org/10.1007/s10551-014-2343-0

Harrison, J. S., & Wicks, A. C. (2013). Stakeholder theory, value, and firm performance. *Business Ethics Quarterly, 23*(1), 97-124. https://doi.org/10.5840/beq20132314

Hekman, D. R., Johnson, S. K., Foo, M., & Yang, W. (2017). Does diversity-valuing behavior result in diminished performance ratings for non-white and female leaders? *Academy of Management Journal, 60*(2), 771-797. https://doi.org/10.5465/amj.2014.0538

Hirokawa, R. Y., & Johnston, D. D. (1989). Toward a general theory of group decision making: Development of an integrated model. *Small Group Research, 20*(4), 500-523. https://doi.org/10.1177/104649648902000408

Horwitz, S. K., & Horwitz, I. B. (2007). The effects of team diversity on team outcomes: A meta-analytic review of team demography. *Journal of Management, 33*(6), 987-1015. https://doi.org/10.1177/0149206307308587

Houkamau, C., & Boxall, P. (2015). Attitudes to other ethnicities among New Zealand workers. *Cross Cultural Management, 22*(3), 431-446. https://doi.org/10.1108/CCM-10-2013-0155

Imenda, S. (2014). Is there a conceptual difference between theoretical and conceptual frameworks? *Journal of Social Sciences, 38*(2), 185–195. https://doi.org/10.1080/09718923.2014.11893249

Jiang, W., Zhao, X., & Ni, J. (2017). The impact of transformational leadership on employee sustainable performance: The mediating role of organizational citizenship behavior. *Sustainability, 9*(9), 1567-1583. https://doi.org/10.3390/su9091567

Kaefer, F., Roper, J., & Sinha, P. (2015). A software-assisted qualitative content analysis of news articles: Example and reflections. *Forum: Qualitative Social Research, 16*(2). https://doi.org/10.17169/fqs-16.2.2123

Kerr, N. L., & Tindale, R. S. (2004). Group performance and decision making. *Annual Review of Psychology, 55*(1), 623-655. https://doi.org/10.1146/annurev.psych.55.090902.142009

Kim, J. Y., Fitzsimons, G. M., & Kay, A. C. (2018). Lean in messages increase attributions of women's responsibility for gender inequality. *Journal of Personality and Social Psychology, 115*(6), 974–1001. https://doi.org/10.1037/pspa0000129

Kirsch, A. (2018). The gender composition of corporate boards: A review and research agenda. *The Leadership Quarterly, 29*(2), 346-364. https://doi.org/10.1016/j.leaqua.2017.06.001

Kluge, I. (2006). Bahá'í ontology, part two: Further explorations. *Lights of Irfan, 7*, 163–200. https://doi.org/http://irfancolloquia.org/60/kluge_ontology

Kohl, K., & Prikladnicki, R. (2018). Perceptions on diversity in Brazilian agile software development teams: A survey. *Proceedings of the 2018 ACM/IEEE 1st International Workshop on Gender Equality in Software Engineering,* 37-40. https://doi.org/10.1145/3195570.3195573

Kulkarni, A., Yoon, I., Pennings, P. S., Okada, K., & Domingo, C. (2018). Promoting Diversity in Computing. *Proceedings of the 23rd Annual ACM Conference on Innovation and Technology in Computer Science Education,* 236-241. https://doi.org/10.1145/3197091.3197145

Kundu, S. C., & Mor, A. (2017). Workforce diversity and organizational performance: A study of IT industry in India. *Employee Relations, 39*(2), 160-183. https://doi.org/10.1108/ER-06-2015-0114

Lamiraud, K., & Vranceanu, R. (2018). Group gender composition and economic decision-making: Evidence from the kallystée business game. *Journal of Economic Behavior and Organization, 145*, 294-305. https://doi.org/10.1016/j.jebo.2017.09.020

Latham, J. R. (2016). *The research canvas: A framework for designing and aligning the "DNA" of your study.* Leadership Plus Design. http://johnlatham.me/researchcanvasbook

London, M., Bear, J. B., Cushenbery, L., & Sherman, G. D. (2019). Leader support for gender equity: Understanding prosocial goal orientation, leadership motivation, and power sharing. *Human Resource Management Review, 29*(3), 418-427. https://doi.org/10.1016/j.hrmr.2018.08.002

Lord, R. G., Day, D. V., Zaccaro, S. J., Avolio, B. J., & Eagly, A. H. (2017). Leadership in applied psychology: Three waves of theory and research. *The Journal of Applied Psychology, 102*(3), 434-451. https://doi.org/10.1037/apl0000089

Macaulay, C. D., Richard, O. C., Peng, M. W., & Hasenhuttl, M. (2018). Alliance network centrality, board composition, and corporate social performance. *Journal of Business Ethics, 151*(4), 997-1008. https://doi.org/10.1007/s10551-017-3566-7

Mazibuko, J. V., & Govender, K. K. (2017). Exploring workplace diversity and organisational effectiveness: A South African exploratory case study. *SA Journal of Human Resource Management, 15*, 1683-7584. https://doi.org/10.4102/sajhrm.v15i0.865

McGee, K. (2018). The influence of gender, and race/ethnicity on advancement in information technology (IT). *Information and Organization, 28*(1), 1-36. https://doi.org/10.1016/j.infoandorg.2017.12.001

Menezes, Á., & Prikladnicki, R. (2018). Diversity in software engineering. *Proceedings of the ACM/IEEE 11th International Workshop on Cooperative and Human Aspects of Software Engineering*, 45-48. https://doi.org/10.1145/3195836.3195857

Meric, I., Er, M., & Gorun, M. (2015). Managing diversity in higher education: USAFA case. *Procedia-Social and Behavioral Sciences, 195*, 72-81. https://doi.org/10.1016/j.sbspro.2015.06.331

Merriam, S. B., & Tisdell, E. J. (2016). *Qualitative research: A guide to design and implementation* (4th ed.). Jossey-Bass.

Montei, M. S., Adams, G. A., & Eggers, L. M. (1996). Validity of scores on the attitudes toward diversity scale (ATDS). *Educational and Psychological Measurement, 56*(2), 293-303. https://doi.org/10.1177/0013164496056002010

Moreno-Gómez, J., Lafuente, E., & Vaillant, Y. (2018). Gender diversity in the board, women's leadership and business performance. Gender in Management: *An International Journal, 33*(2), 104-122. https://doi.org/10.1108/GM-05-2017-0058

Nakui, T., Paulus, P. B., & van der Zee, Karen I. (2011). The role of attitudes in reactions toward diversity in workgroups. *Journal of Applied Social Psychology, 41*(10), 2327-2351. https://doi.org/10.1111/j.1559-1816.2011.00818.x

Nikolova, H., & Lamberton, C. (2016). Men and the middle: Gender differences in dyadic compromise effects. *Journal of Consumer Research, 43*(3), 355-371. https://doi.org/10.1093/jcr/ucw035

Ohunakin, F., Adeniji, A., Ogunnaike, O. O., Igbadume, F., & Akintayo, D. I. (2019). The effects of diversity management and inclusion on organisational outcomes: A case of multinational corporation. *Business: Theory and Practice, 20*, 93-102. https://doi.org/10.3846/btp.2019.09

O'Reilly, M., & Parker, N. (2013). 'Unsatisfactory saturation': A critical exploration of the notion of saturated sample sizes in qualitative research. *Qualitative Research, 13*(2), 190–197. https://doi.org/10.1177/1468794112446106

Palmatier, R. W., Houston, M. B., & Hulland, J. (2018). Review articles: Purpose, process, and structure. *Journal of the Academy of Marketing Science, 46*(1), 1-5. https://doi.org/10.1007/s11747-017-0563-4

Parmar, B. L., Freeman, R. E., Harrison, J. S., Wicks, A. C., Purnell, L., & de Colle, S. (2010). Stakeholder theory: The state of the art. *The Academy of Management Annals, 4*(1), 403-445. https://doi.org/10.5465/19416520.2010.495581

Patton, M. Q. (2014). *Qualitative research & evaluation methods: Integrating theory and practice.* Sage Publications.

Percy, W.H., Kostere, K., & Kostere, S. (2015). Generic qualitative research in psychology. *The Qualitative Report, 20*(2), 76-85 https://doi.org/10.46743/2160-3715/2015.2097

Powell, G. N. (1998). Reinforcing and extending today's organizations: The simultaneous pursuit of person-organization fit and diversity. *Organizational Dynamics, 26*(3), 50-61. https://doi.org/10.1016/S0090-2616(98)90014-6

Quadlin, N. (2018). The mark of a woman's record: Gender and academic performance in hiring. *American Sociological Review, 83*(2), 331-360. https://doi.org/10.1177/0003122418762291

Rao, K., & Tilt, C. (2016). Board composition and corporate social responsibility: The role of diversity, gender, strategy and decision making. *Journal of Business Ethics, 138*(2), 327-347. https://doi.org/10.1007/s10551-015-2613-5

Rocco, T. S., & Plakhotnik, M. S. (2009). Literature Reviews, Conceptual Frameworks, and Theoretical Frameworks: Terms, Functions, and Distinctions. *Human Resource Development Review, 8*(1), 120–130. https://doi.org/10.1177/1534484309332617

Room, A., & Brewer, E. C. (Eds.). (2009). Seminal work. In *Brewer's dictionary of modern phrase and fable* (2nd ed.). Cassell.

Sager, K. L., & Gastil, J. (1999). Reaching consensus on consensus: A study of the relationships between individual decision-making styles and use of the consensus decision rule. *Communication Quarterly, 47*(1), 67-79. https://doi.org/10.1080/01463379909370124

Salmona, M., & Kaczynski, D. (2016). Don't blame the software: Using qualitative data analysis software successfully in doctoral research. *Qualitative Social Research, 17*(3). https://doi.org/10.17169/fqs-17.3.2505

Saunders, B., Kitzinger, J., & Kitzinger, C. (2014). Anonymising interview data: Challenges and compromise in practice. *Qualitative Research, 15*(5), 616–632. https://doi.org/10.1177/1468794114550439

Scharmer, C. O., & Kaeufer, K. (2010). In front of the blank canvas: Sensing emerging futures. *The Journal of Business Strategy, 31*(4), 21-29. https://doi.org/10.1108/02756661011055159

Schein, E. H. (2002). Models and tools for stability and change in human systems. *Reflections: The SoL Journal, 4*(2), 34–46. https://doi.org/10.1162/152417302762251327

Senge, P.M. (2006). *The fifth discipline: The art & practice of the learning organization.* Broadway Business.

Senge, P. M., & Fulmer, R. M. (1993). Simulations, systems thinking and anticipatory learning. *The Journal of Management Development, 12*(6), 21. https://doi.org/10.1108/02621719410050228

Tajfel, H. (1974). Social identity and intergroup behavior. *Social Science Information, 13*(2), 65-93. https://doi.org/10.1177/053901847401300204

Thomas, D. A., & Ely, R. J. (1996). Making differences matter: A new paradigm for managing diversity. *Harvard Business Review, 74*(5), 79-90.

Trochim, W. M. K. (2006). *Qualitative validity.* Research Methods Knowledge Base. http://www.socialresearchmethods.net/kb/qualval.htm

Tsui, A. S., Egan, T. D., & O'Reilly, C. A. (1992). Being different: Relational demography and organizational attachment. *Administrative Science Quarterly, 37*(4), 549-579. https://doi.org/10.2307/2393472

von Alberti-Alhtaybat, L., & Al-Htaybat, K. (2010). Qualitative accounting research: An account of Glaser's grounded theory. *Qualitative Research in Accounting and Management, 7*(2), 208–226. https://doi.org/10.1108/11766091011050868

von Bertalanffy, L. (1957). General system theory. In R. W. Taylor (Ed.), *Life, language, law: Essays in honor of Arthur F. Bentley* (pp. 58–78). The Antioch Press.

von Bertalanffy, L. (1972). The history and status of general systems theory. *The Academy of Management Journal, 15*(4), 407-426. https://doi.org/10.2307/255139

Vroom, V. H., & Jago, A. G. (1974). Decision making as a social process: Normative and descriptive models of leader behavior. *Decision Sciences, 5*(4), 743-769. https://doi.org/10.1111/j.1540-5915.1974.tb00651.x

Willgens, A., Cooper, R., Jadotte, D., Lilyea, B., Langtiw, C., & Obenchain-Leeson, A. (2016). How to enhance qualitative research appraisal: Development of the methodological congruence instrument. *Qualitative Report, 21*(12), 2380. https://doi.org/10.46743/2160-3715/2016.2361

Williams, M., & Moser, T. (2019). The art of coding and thematic exploration in qualitative research. *International Management Review, 15*(1), 45-72.

Williams, M. P. (2003). *Knowledge management: An evolving professional discipline* [Doctoral book, Fordham University]. ProQuest Dissertations Publishing. https://www.proquest.com/docview/305327901?pq

Wondrak, M., & Segert, A. (2015). Using the diversity impact navigator to move from interventions towards diversity management strategies. *Journal of Intellectual Capital, 16*(1), 239-254. https://doi.org/10.1108/JIC-12-2013-0117

Wynn, A. T. (2020). Pathways toward change: Ideologies and gender equality in a Silicon Valley technology company. *Gender & Society, 34*(1), 106-130. https://doi.org/10.1177/0891243219876271

Yu, H. H. (2018). Gender and public agency hiring: An exploratory analysis of recruitment practices in federal law enforcement. *Public Personnel Management, 47*(3), 247-264. https://doi.org/10.1177/0091026018767473

Yukl, G. (1999). An evaluation of conceptual weaknesses in transformational and charismatic leadership theories. *The Leadership Quarterly, 10*(2), 285-305. https://doi.org/10.1016/S1048-9843(99)00013-2